Springer Series in Materials Science

Volume 232

The Springer Series in Materials Science covers the complete spectrum of materials physics, including fundamental principles, physical properties, materials theory and design. Recognizing the increasing importance of materials science in future device technologies, the book titles in this series reflect the state-of-the-art in understanding and controlling the structure and properties of all important classes of materials.

More information about this series at http://www.springer.com/series/856

Andreas Trügler

Optical Properties of Metallic Nanoparticles

Basic Principles and Simulation

 Springer

Andreas Trügler
Graz, Austria

ISSN 0933-033X   ISSN 2196-2812   (electronic)
Springer Series in Materials Science
ISBN 978-3-319-79730-4   ISBN 978-3-319-25074-8   (eBook)
DOI 10.1007/978-3-319-25074-8

Springer Cham Heidelberg New York Dordrecht London
© Springer International Publishing Switzerland 2016
Softcover reprint of the hardcover 1st edition 2016

Printed on acid-free paper

Springer International Publishing AG Switzerland is part of Springer Science+Business Media
(www.springer.com)

Dedicated to Eva and
the memory of my grandfather Fritz

Preface

The roots of this book date back to a discussion I had with my colleague Ulrich Hohenester. Several years ago he gave me an overview of the research topics he was working on at that time, I chose plasmonics, and Ulrich became my PhD advisor and friend. I owe sincere thanks to him for the countless discussions we had, for conveying his passion for physics, and for his subtle way of encouragement. Ulrich has a great talent to bring clearness into intricate problems (be it physics or anything else) and with these pages I seek to pass on some of the insights that accompanied our collaboration. The core of this book is essentially my PhD thesis updated with a couple of new chapters, some interspersed remarks and explanations, several new findings, and hot off the press references. It is meant as an introduction to the fascinating world of physics at the nanoscale with a focus on simulations and the theoretical aspects of plasmonics. The extensive references at the end of each chapter should allow a continuation of the search for knowledge whenever the scope of this book reaches a limit. The mathematical prerequisites probably comprise a bachelor or master's degree in physics, but besides the formal parts this book also contains over 100 figures and schematics where I tried to pin down the actual events and principles in a graphical way without any equations.

If you look back in history, throughout the centuries the progress in physics has largely been dominated by experiments. There are some exceptions of course, but at least the entanglement between theory and experiment has always been essential in understanding our nature. However ingenious a theory may have been, it remained a chimera until there was experimental verification. Because of the continuing specialization in physics also the multidisciplinarity of a research field becomes more and more important, and as a theoretical physicist I am therefore very positive about the manifold possibilities for collaborations with experimentalists and other scientists in my work. This book is also strongly influenced by such collaborations, and I want to express my gratitude for that. In alphabetical order I sincerely thank Franz Aussenegg, Martin Belitsch, Harald Ditlbacher, Marija Gašparić, Christian Gruber, Andreas Hohenau, Daniel Koller, Joachim Krenn, Markus Krug, Verena Leitgeb, Alfred Leitner, Gernot Schaffernak, Franz Schmidt, Jean-Claude Tinguely, Pamina Winkler and all other (current and several former) members of the Optical

Nanotechnology/Nano-Optics group in Graz for most helpful discussions and the fruitful teamwork.

Furthermore and following the thematic order of this book, I want to thank the research group of Carsten Sönnichsen at the Institute for Physical Chemistry in Mainz, where Jan Becker and others did a marvelous job in putting the first versions of our MNPBEM toolbox to the acid test; the group of Ferdinand Hofer at the Austrian Centre for Electron Microscopy and Nanoanalysis (FELMI) for their ongoing cooperation on EELS and plasmon tomography; the group of Alfred Leitenstorfer from the Lehrstuhl für Moderne Optik und Quantenelektronik (especially Tobias Hanke for his collaboration on nonlinear optics) as well as the group of Rudolf Bratschitsch at the University of Münster, where I was working with Johannes Kern on effects of very thin substrate layers, and last but not least Javier García de Abajo and his group for a very pleasant research stay during which I started working on nonlocal plasmonics.

Special thanks also go to Polly Cassidy for her marvelous job as lector of my manuscript as well as to the members of my own research group: Anton Hörl, Dario Knebl, Robert Schütky, Jürgen Waxenegger and all others, who contribute to the pleasant atmosphere at our institute.

Finally I owe my deepest gratitude to all my friends and family, without their support and encouragement this book would not have been possible.

Graz, Austria   Andreas Trügler
2015

Contents

Acronyms

Table of symbols

$\tilde{A}$	Complex $3N \times 3N$ matrix
A	Vector potential of Maxwell's theory
B	Magnetic field (historically also magnetic induction), $B = \mu H$
C_{abs}	Absorption cross section
C_{ext}	Extinction cross section
C_{sca}	Scattering cross section
C	Closed contour of an area or capacitance of a capacitor
Δh	Scaling of stochastic height variations h
Δk_x	Uncertainty of the momentum of a microscopic particle in the spatial x-direction
Δx	Uncertainty in the spatial position of a microscopic particle
D	Dielectric displacement, $D = \varepsilon E$
E_F	Fermi energy, $E_F \approx 5.53$ eV for gold or silver
E_S	Enhancement factor for SERS
$\tilde{E}$	$3N$-dimensional (complex) vector of the electric field at each lattice site
E	Energy of a certain state
E	Electric field
F_L	Lorentz force
F	Surface derivative of Green function G expressed as a matrix
$\mathcal{F}^{-1}$	Inverse Fourier transform
$\mathcal{F}$	Fourier transform
F	Surface derivative of Green function G
G_2	Second order autocorrelation function
G_3	Third order autocorrelation function
Γ	Homogeneous linewidth (FWHM) of a spectral resonance or electron energy-loss probability
$\mathcal{G}$	Susceptibility kernel, Fourier transform of χ_e
G	Green function, same symbol for quasistatic and retarded case

H	Magnetic field, $H = \frac{1}{\mu}B$
Λ	Matrix containing the dielectric information of a quasistatic problem
L_i	Geometrical shape factor for ellipsoidal particles
L	Shape factor, characteristic length scale of a structure or inductance of a magnetic coil
$\mathcal{M}$	Vector harmonic, $\mathcal{M} = \nabla \times (r\psi)$
M	Magnetic moment per unit volume, $M = \chi_m H$
NA	Numerical aperture of an optical system
$\mathcal{N}$	Vector harmonic, $\mathcal{N} = \frac{1}{k}\nabla \times \mathcal{M}$
N_ω	Number of calculated frequencies
N	Normalization factor ($N \in \mathbb{R}$) or total number of entities ($N \in \mathbb{N}$), e.g. lattice sites or number of phases
Ω	Specific region, e.g. a metallic nanoparticle or a sphere segment
$\mathcal{O}$	Order of a Taylor series (Landau notation)
$\mathcal{P}$	Cauchy principal value
$\tilde{P}$	$3N$-dimensional (complex) vector of the dipole polarization at each lattice site
ϕ	Scalar potential
$P_\alpha^{(L)}$	αth component of the linear dipole moment per unit volume, $\alpha \in \{x, y, z\}$
P_l^m	Associated Legendre functions of the first kind of degree l and order m
P	Dipole moment per unit volume (polarization), $P = \varepsilon_0 \chi_e E$
R_F	Förster radius, typically $2\,\text{nm} < R_F < 9\,\text{nm}$.
$\mathcal{R}$	Radon transform
R	Radius of a sphere or circle or the distance between a donor and an acceptor molecule
S	Poynting vector (directional energy flux density), $S = E \times H$, $\langle S \rangle = \frac{1}{2}\mathfrak{Re}\{E \times H\}$
S	Sensitivity, often simply denoted as $\Delta\lambda/\text{RIU}$, can be expressed in wavelength (S_λ) or energy (S_E) units
T	Time constant
V_atom	Volume of a metal atom
V_uc	Unit cell volume of a crystal structure
V	Volume of a nanoparticle
Y_{lm}	Spherical harmonics of degree l and order m
α	Polarizability tensor
α	Polarizability (e.g. of a spheroid) or fine-structure constant $\alpha = \frac{1}{4\pi\varepsilon_0}\frac{e^2}{\hbar c} \approx \frac{1}{137}$
a_l	Mie scattering coefficient for mode l
a	Lattice constant, spatial dimension of the unit cell in a crystal
β	Nonlocal parameter within the hydrodynamical model, $\beta = \sqrt{3/5}\,v_F$
b_l	Mie scattering coefficient for mode l

χ_d	Electric susceptibility of the Drude model
χ_e	Electric susceptibility
$\chi^{(i)}$	Susceptibility tensor of rank $(i+1)$
χ_m	Magnetic susceptibility
c_l	Mie coefficient for inside field and mode l
c	Speed of light, in vacuum: $299\,792\,458\ \mathrm{m/s} \approx 300\ \mathrm{nm/fs}$
ΔE	Energy loss in EELS
δF	Small deviation of F due to surface roughness
$\partial \Omega$	Boundary of a region Ω, e.g. the surface of a nanoparticle
δ	Plasmon propagation length
d_l	Mie coefficient for inside field and mode l
$\boldsymbol{d}$	Dipole moment of a two level system, e.g. a molecule
$d\Omega$	Element of the solid angle, in spherical polar coordinates: $d\Omega = \sin(\theta)d\theta d\varphi$
$\hat{\boldsymbol{e}}$	Unit vector, for example along the axis of an orthogonal coordinate system
ε_0	Vacuum permittivity or electric constant, $\varepsilon_0 \equiv {}^1/_{\mu_0 c^2} = {}^{e^2}/_{4\pi\alpha\hbar c} = 8.854187817 \times 10^{-12}\ \mathrm{F/m}\ \left(= {}^{\mathrm{C}}/_{\mathrm{Vm}} = {}^{\mathrm{As}}/_{\mathrm{Vm}} = {}^{\mathrm{A^2\,s^4}}/_{\mathrm{kg\,m^3}}\right)$
e	Electron charge, $e = 1.60217657 \times 10^{-19}\ \mathrm{C}$
f	Real or complex-valued function
γ	Decay rate
g	coupling strength between a quantum emitter and a cavity mode
$\bar{\boldsymbol{h}}$	Electromagnetic surface current density, $[\bar{\boldsymbol{h}}] = {}^{\mathrm{A}}/_{\mathrm{m}}$
$\hbar$	Reduced Planck constant or Dirac constant, $\hbar = \frac{h}{2\pi} = 6.58211928 \times 10^{-16}\ \mathrm{eV\,s} = 1.054571726 \times 10^{-34}\ \mathrm{J\,s}$
$\boldsymbol{h}$	Artificial surface current density, $[\boldsymbol{h}] = {}^{\mathrm{Vs}}/_{\mathrm{m^2}}$
h	Stochastic height variations to model surface roughness
j	Subscript index referring to a dielectric medium j or a general counting index, $j \in \mathbb{N}$
$\boldsymbol{j}$	Electromagnetic current density, $[\boldsymbol{j}] = {}^{\mathrm{A}}/_{\mathrm{m^2}}$
k_B	Boltzmann constant, $k_\mathrm{B} = 1.381 \times 10^{-23}\ {}^{\mathrm{J}}/_{\mathrm{K}} = 8.617 \times 10^{-5}\ {}^{\mathrm{eV}}/_{\mathrm{K}}$
$\boldsymbol{k}$	Wave vector of light, $|\boldsymbol{k}| = k = \frac{2\pi}{\lambda} = n\frac{\omega}{c} = \omega\sqrt{\mu\varepsilon}$
k_j	Wave number in medium j
$\tilde{k}$	Imaginary part of complex refractive index $n = \tilde{n} + \mathrm{i}\tilde{k}$
λ_k	Eigenenergy of mode k of matrix $\boldsymbol{F}$
λ	Wavelength of light, $\lambda = \frac{2\pi}{k} = \frac{c}{n}\frac{2\pi}{\omega}$
$\triangle$	Laplace operator, $\triangle \equiv \boldsymbol{\nabla}^2$
m_e	Mass of the free electron, $m_e = 9.10938291 \times 10^{-31}\ \mathrm{kg}$
m_γ	Photon mass, $m_\gamma = 0$ (experiment: $m_\gamma < 4 \times 10^{-51}\ \mathrm{kg}$)
μ_0	Vacuum permeability or magnetic constant, $\mu_0 \equiv 4\pi \times 10^{-7}\ {}^{\mathrm{Vs}}/_{\mathrm{Am}}$
μ	Magnetic permeability, $\mu \approx \mu_0$ for metals at optical wavelengths
n_e	Electron density

$\hat{n}$	Unit vector (for example pointing in the scattering direction)
$\tilde{n}$	Real part of complex refractive index $n = \tilde{n} + i\tilde{k}$
ν	Frequency of a photon, $\nu = \omega/2\pi = c/\lambda$, or vibrational energy level of a molecule
n	Refractive index of a dielectric medium ($n = c_{vac}/c_{med}$), for metals $n = \sqrt{\mu\varepsilon/\mu_0\varepsilon_0} \in \mathbb{C}$ and $\mu \approx \mu_0$ at optical frequencies
ω_{mp}	Magnetic plasma frequency
ω_p	(Bulk) plasma frequency of the conduction electrons of a metal ($\hbar\omega_p = \hbar\sqrt{n_e e^2/\varepsilon_0 m_e} = 9.07$ eV for gold or silver)
ω	Angular frequency of a photon, $\omega = 2\pi\nu = 2\pi\frac{c}{\lambda}$
ϕ_{rnd}	Stochastic phase factors
ϕ_{ext}	Potential of an external excitation like a plane wave or a dipole
ψ	Generating function for the vector harmonics $\mathcal{N}$ and $\mathcal{M}$, $\psi = \psi(r, \theta, \varphi)$
p	Norm of the momentum $\boldsymbol{p}$ of a particle
q	Electromagnetic charge
$\boldsymbol{r}$	Spatial position vector, $\boldsymbol{r} \in \mathbb{R}^3$
$\boldsymbol{r}_e$	Position of the electron beam
r	Fresnel reflection coefficient
ρ	Radius in a polar coordinate system
r_s	Electron gas parameter
r	Norm of spatial position vector, radial component of spherical polar coordinates
$\boldsymbol{s}$	Spatial position vector on a surface $\partial\Omega$
σ_k^L	Left eigenvector of $\boldsymbol{F}$, can be interpreted as the kth surface plasmon eigenmode
σ_k^R	Right eigenvector of $\boldsymbol{F}$, can be interpreted as the kth surface plasmon eigenmode
$\bar{\sigma}$	Surface charge density, $[\bar{\sigma}] = C/m^2$
σ	Artificial surface charge density, $[\sigma] = V/m$
τ	Dephasing time of a surface plasmon polariton (typically $\tau < 10$ fs) or time variable
t	Fresnel transmission coefficient
$\hat{t}$	Tangential unit vector
θ	Inclination angle of spherical polar coordinates or angle of incidence on a planar interface
t	Time variable, $t \in \mathbb{R}$
u	Energy density of an electromagnetic wave, in general $u = \frac{1}{2}(\boldsymbol{E}\cdot\boldsymbol{D} + \boldsymbol{B}\cdot\boldsymbol{H})$, in vacuum $u = \frac{1}{2}(\varepsilon_0\boldsymbol{E}^2 + \mu_0\boldsymbol{H}^2)$
v_F	Fermi velocity, $v_F \approx 1.4$ nm/fs for gold or silver
ε	Dielectric description of a material, may be a constant or a frequency dependent (complex) function

φ	Angle in a polar coordinate system
ϱ	External or free charge distribution of Maxwell's theory
ς	Variance of height fluctuations in a Gaussian autocorrelation function
$\boldsymbol{v}$	Velocity of a point charge
x	Real-valued variable or point on the real axis
y	Real-valued variable, second Cartesian coordinate
ζ	Skin depth of evanescent field
z_l	Substitute for any of the four spherical Bessel functions j_l, y_l, $h_l^{(1)}$, or $h_l^{(2)}$
z	Complex number $z = z_1 + \mathrm{i}\, z_2 \in \mathbb{C}$ or third Cartesian coordinate

Part I
Introduction and Basic Principles

Chapter 1
Prologue

> *The electron is a theory we use; it is so useful in understanding*
> *the way nature works that we can almost call it real.*
>
> RICHARD FEYNMAN

Much has happened since the struggle between the devotees of the undulatory and corpuscular theory of light. For millennia we have been fascinated by optical phenomena and the groundbreaking works of many brilliant scientists allowed deep insights into the question *of what holds the world together in its inmost folds.* Step by step we gain more understanding of what light actually is [1]. In particular, the interaction of light with matter is a treasure trove for new applications and a demanding criterion for the underlying physical theories.

1.1 The Glamour of Plasmonics

Half a century ago, Richard Feynman[1] was already aware of the fact that *there is plenty of room at the bottom* [2] and he invited his listeners to open up a new field of physics. In the 1970s the term *nanotechnology* was formed [3, 4] and remarkable progress and new discoveries in the "nanoworld" followed–often resulting in a Nobel Prize for the respective scientists.[2]

The study of optical phenomena related to the electromagnetic response of metals, which is the topic of this book, led to the development of an emerging and fast growing research field called *plasmonics* [5]–named after the electron density waves that propagate along the interface of a metal and a dielectric like the ripples that spread across water after throwing a stone across its surface [6].

In particular, the enormous progress in the fabrication and manipulation methods of nanometer-sized objects in the last decades have allowed us to enter into the

---

[1]Born 11th May 1918 in New York City; † 15th February 1988 in Los Angeles, California. Nobel Prize in Physics 1965.

[2]For example the invention of the scanning tunneling microscope, the discovery of fullerenes, the development of fluorescence microscopy, etc.

© Springer International Publishing Switzerland 2016
A. Trügler, *Optical Properties of Metallic Nanoparticles*, Springer Series in Materials Science 232, DOI 10.1007/978-3-319-25074-8_1

fascinating world at the length scale of molecules and DNA strands. There are certain *promises of plasmonics* [6] that are responsible for the current boom in this research field, such as the prediction of new superfast computer chips [6], new possibilities to treat cancer [7–10], ultrasensitive molecular detectors [11–13] or the ability to make things invisible with negative-refraction materials [14–16]. All this is possible because plasmonics builds a bridge between two different length scales by confining light on sub-wavelength volumes. The building bricks of this arch are metallic nanoparticles (MNPs) and colloids, or thin metal films in the case of plasmonic waveguides. The focus of this book lies on the description of metallic nanoparticles (see Fig. 1.1)–their optical properties, how they influence and interact with their surroundings, and how we can make these events visible although the involved structural sizes are much below the wavelength of light. Besides those already mentioned, the capability to manipulate and control light on the nanometer scale opens up a plethora of further possible applications [17, 18], as diverse as data storage [19], optical data processing [20, 21], quantum optics [22–24], optoelectronics [25, 26], photovoltaics [27], or quantum information processing.

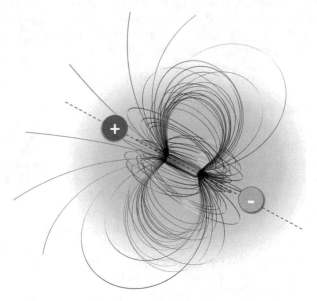

Fig. 1.1 Electric field lines of a gold nanorod ($15 \times 45 \times 8\,\text{nm}^3$) at the resonance wavelength of 852 nm. The corresponding surface charge is plotted on the boundary of the nanorod and shows a dipolar distribution

1.2 Scope of This Book

This book is divided into four main parts. The first one aims at providing a general introduction to the topic as well as a detailed description of the underlying theoretical concepts. At the beginning of Chap. 2 a short historical overview of plasmons is given–from their first observation to their modern perception. This synopsis is then followed by a discussion of surface plasmon polaritons and the rest of the chapter is dedicated to different plasmonic properties and how they can be engineered and exploited. Chapter 3 represents the formal section of this book where the mathematical description of metallic nanoparticles is formulated.

The second part with Chap. 4 involves an introduction and discussion of different numerical approaches to solve the previously derivated equations. Although several simulation methods are briefly discussed, the main focus lies on the boundary element method.

With these numerical and theoretical tools at hand, we can then dig into various plasmonic problems and applications in the third part. Chapter 5 is dedicated to the question of how to picture a plasmon, Chaps. 6 and 7 cover the influence of surface roughness and nonlinear optical effects of metallic nanoparticles, respectively. In Chap. 8 the case of very small particles and nonlocal effects that come along with spatial dependent dielectric functions is discussed. Chapter 9 is devoted to so-called metamaterials–artificial new materials with bizarre optical properties.

The last part of this book contains supplementary material such as appendices, a list of figures and tables as well as a cross-reference index. Several important relations are highlighted by a box throughout this book and collected in the list of equations. Appendix A contains additional derivations and explanations. Finally, in Appendix B the MATLAB® source code for the simulation of spherical nanoparticles is attached.

An extensive list of references and further reading is provided after each chapter and to make work easier with the cited papers and books, they have been cross-referenced with online links in the e-book version; the symbol ▤ in the bibliography constitutes a direct link to pdf-files whenever possible at the time of writing these pages.

1.3   Measurement Units

The question of a suitable unit system often comes as little bit of a nuisance, especially if you ask a theoretical physicist. It is where we have to bring our equations
down to earth, where our mathematical language should cough up actual numbers
and where we have to face the reality of experimental outcomes. In plasmonics it
is quite common to use Gaussian units, at least among theoreticians. They bring
the advantage that the electromagnetic fields obtain the same dimensions, that the
vacuum permittivity ε_0 and the vacuum permeability μ_0 vanish, and they seem to be
a more natural description of classical field theory and relativistic electromagnetism.
The *système international d'unités*[3] (or abbreviated SI unit system) on the other
hand, is the most common unit system, in science as well as everywhere else. In
the last years many of the excellent books about Maxwell's equations or plasmonics
have converted to SI units (for example the fabulous standard work of Jackson [28],
see his preface to the third edition) or have a priori been developed in SI units
(e.g. the excellent book of Novotny and Hecht [5] or the comprehensive work of
Chew [29]). With this regard and also because of the multidisciplinarity of plasmonics I also decided to use SI units throughout this book. Here is a list of units of
the most important electromagnetic quantities appearing in the following chapters.

Electric field, dielectric displacement and polarization:

$$[\boldsymbol{E}] = \frac{\mathrm{V}}{\mathrm{m}}, \qquad\qquad [\boldsymbol{D}] = \frac{\mathrm{C}}{\mathrm{m}^2}, \qquad\qquad [\boldsymbol{P}] = \frac{\mathrm{C}}{\mathrm{m}^2}.$$

Magnetic field and magnetization:

$$[\boldsymbol{B}] = T = \frac{\mathrm{V s}}{\mathrm{m}^2}, \qquad\qquad [\boldsymbol{H}] = \frac{\mathrm{A}}{\mathrm{m}}, \qquad\qquad [\boldsymbol{M}] = \frac{\mathrm{A}}{\mathrm{m}}.$$

Charge density, current density and unit charge

$$[\rho] = \frac{\mathrm{C}}{\mathrm{m}^3}, \qquad\qquad [\boldsymbol{j}] = \frac{\mathrm{A}}{\mathrm{m}^2}, \qquad\qquad [q] = C = \mathrm{As}.$$

Electromagnetic potentials and nabla operator:

$$[\phi] = \mathrm{V}, \qquad\qquad [\boldsymbol{A}] = \frac{\mathrm{V s}}{\mathrm{m}}, \qquad\qquad [\boldsymbol{\nabla}] = \frac{1}{\mathrm{m}}.$$

---

[3]As the modern version of the metric system, it is an achievement of the French Revolution and
therefore still retains a French name.

The SI units in this list are Volt (V), Coulomb (C), meters (m), seconds (s), and Ampere (A).

The vacuum permittivity, vacuum permeability, and consequently the speed of light yield

$$\left.\begin{array}{l} \varepsilon_0 = 8.854187817 \times 10^{-12} \dfrac{\mathrm{F}}{\mathrm{m}} \left(= \dfrac{\mathrm{As}}{\mathrm{Vm}} \right) \\[3ex] \mu_0 \equiv 4\pi \times 10^{-7} \dfrac{\mathrm{H}}{\mathrm{m}} \left(= \dfrac{\mathrm{Vs}}{\mathrm{Am}} \right) \end{array}\right\} \quad [c] = \left[\dfrac{1}{\sqrt{\mu_0 \varepsilon_0}} \right] = \dfrac{\mathrm{m}}{\mathrm{s}},$$

where the derived SI units Henry (H) and Farad (F) have been used. For linear response we have for the material parameters

$$\varepsilon = \varepsilon_0 (1 + \chi_e), \qquad \mu = \mu_0 (1 + \chi_m),$$

with the dimensionless electric and magnetic susceptibilities χ_e and χ_m, respectively. For nonlinear response (see Chap. 7) the higher order magnetic susceptibility remains dimensionless, whereas for the electric susceptibility of rank n we get

$$\left[\chi_e^{(n)} \right] = \left(\frac{\mathrm{m}}{\mathrm{V}} \right)^{n-1} .$$

In the earlier literature usually the electrostatic units (esus) of the susceptibilities have been tabulated, the conversion to SI values is straight forward:

$$\frac{\chi_e^{(n)}[\mathrm{SI}]}{\chi_e^{(n)}[\mathrm{esu}]} = \frac{4\pi}{(10^{-4}c)^{n-1}} .$$

The surface charge and surface current density are expressed in $\mathrm{C/m^2}$ and $\mathrm{A/m}$, respectively. To be consistent with the original derivation of the boundary element method a redefinition is necessary in Chap. 3, which changes their units to $\mathrm{V/m}$ and $\mathrm{Vs/m^2}$, respectively. The quantities plotted in the figures are almost exclusively expressed in femtoseconds (fs), electron Volts (eV), or nanometers (nm)–the common dimensions in plasmonics.

References

1. J. Heber, A. Trabesinger, Nature milestones: photons. Nat. Mater. **9**, S5–S20 (2010).
2. R. Feynman, There is plenty of room at the bottom (talk transcript). Caltech Eng. Sci. **23**(5), 22–36 (1960).
3. N. Taniguchi (ed.), *On the Basic Concept of Nano-Technology, Proceedings of the International Conference on Production Engineering*, Part II (Japan Society of Precision Engineering, Tokyo, 1974)
4. K.E. Drexler, *Molecular Machinery and Manufacturing with Applications to Computation.* Ph.D. thesis, Massachusetts Institute of Technology (MIT), Cambridge, 1991.
5. L. Novotny, B. Hecht, *Principles of Nano-Optics*, 2nd edn. (Cambridge University Press, Cambridge, 2012). ISBN 978-1107005464
6. H. Atwater, The promise of plasmonics. Sci. Am. **296**(4), 56 (2007).
7. I.H. El-Sayed, X. Huang, M.A. El-Sayed, Selective laser photo-thermal therapy of epithelial carcinoma using anti-EGFR antibody conjugated gold nanoparticles. Cancer Lett. **239**, 129–135 (2006).
8. J.M. Stern, J. Stanfield, W. Kabbani, J.-T. Hsieh, J.A. Cadeddu, Selective prostate cancer thermal ablation with laser activated gold nanoshells. The J. Urol. **179**(2), 748–753 (2008).
9. J.Z. Zhang, Biomedical applications of shape-controlled plasmonic nanostructures: a case study of hollow gold nanospheres for photothermal ablation therapy of cancer. J. Phys. Chem. Lett. **1**(4), 686–695 (2010).
10. G.A. Sotiriou, F. Starsich, A. Dasargyri, M.C. Wurnig, F. Krumeich, A. Boss, J.-C. Leroux, S.E. Pratsinis, Photothermal killing of cancer cells by the controlled plasmonic coupling of silica-coated Au/Fe$_2$O$_3$ nanoaggregates. Adv. Funct. Mater. **24**(19), 2818–2827 (2014).
11. J.N. Anker, W.P. Hall, O. Lyandres, N.C. Shah, J. Zhao, R.P.V. Duyne, Biosensing with plasmonic nanosensors. Nat. Mater. **7**, 442 (2008).
12. A. Haes, R.P.V. Duyne, A nanoscale optical biosensor: sensitivity and selectivity of an approach based on the localized surface plasmon resonance spectroscopy of triangular silver nanoparticles. J. Am. Chem. Soc. **124**, 10596 (2002).
13. P. Fortina, L.J. Kricka, D.J. Graves, J. Park, T. Hyslop, F. Tam, N. Halas, S. Surrey, S.A. Waldman, Applications of nanoparticles to diagnostics and therapeutics in colorectal silver nanoparticles. J. Am. Chem. Soc. **124**, 10596 (2007).
14. J.B. Pendry, Negative refraction makes a perfect lens. Phys. Rev. Lett. **85**, 3966–3969 (2000).
15. H.J. Lezec, J.A. Dionne, H.A. Atwater, Negative refraction at visible frequencies. Science **316**, 5823 (2007).
16. N.I. Zheludev, A roadmap for metamaterials. Opt. Photonics News **22**, 30–35 (2011).
17. A. Polman, Plasmonics applied. Science **322**(5903), 868–869 (2008).
18. T. Nagao, G. Han, C. Hoang, J.-S. Wi, A. Pucci, D. Weber, F. Neubrech, V.M. Silkin, D. Enders, O. Saito, M. Rana, Plasmons in nanoscale and atomic-scale systems. Sci. Technol. Adv. Mater. **11**(5), 054506 (2010).
19. P. Zijlstra, J.M. Chon, M. Gu, Five-dimensional optical recording mediated by surface plasmons in gold nanorods. Nature **459**, 410–413 (2009).
20. B. Lamprecht, J.R. Krenn, G. Schider, H. Ditlbacher, M. Salerno, N. Felidj, A. Leitner, F.R. Aussenegg, J.C. Weeber, Surface plasmon propagation in microscale metal stripes. Appl. Phys. Lett. **79**, 51 (2001).
21. S.I. Bozhevolnyi, V.S. Volkov, E. Deveaux, J.Y. Laluet, T.W. Ebbensen, Channel plasmon subwavelength waveguide components including interferometers and ring resonators. Nature **440**, 509 (2006).

22. A. Akimov, A. Mukherjee, C.L. Yu, D.E. Chang, A.S. Zibrov, P.R. Hemmer, H. Park, M.D. Lukin, Generation of single optical plasmons in metallic nanowires coupled to quantum dots. Nature **450**, 402–406 (2007).
23. R. Kolesov, B. Grotz, G. Balasubramanian, R.J. Stöhr, A.A.L. Nicolet, P.R. Hemmer, F. Jelezko, J. Wachtrup, Wave-particle duality of single surface plasmon polaritons. Nat. Phys. **5**, 470–474 (2009).
24. R. Loudon, *The Quantum Theory of Light* (Oxford University Press, New York, 2000). ISBN 978-0198501763
25. D.M. Koller, A. Hohenau, H. Ditlbacher, N. Galler, F. Reil, F.R. Aussenegg, A. Leitner, E.J.W. List, J.R. Krenn, Organic plasmon-emitting diode. Nat. Phot. **2**, 684–687 (2008).
26. A.L. Falk, F.H.L. Koppens, C.L. Yu, K. Kang, N. de Leon Snapp, A.V. Akimov, M.-H. Jo, M.D. Lukin, H. Park, Near-field electrical detection of optical plasmons and single-plasmon sources. Nat. Phys. **5**, 475–479 (2009).
27. V.E. Ferry, L.A. Sweatlock, D. Pacifici, H.A. Atwater, Plasmonic Nanostructure design for efficient light coupling into solar cells. Nano Lett. **8**, 4391 (2008).
28. J.D. Jackson, *Classical Electrodynamics* (Wiley, New York, 1962). ISBN 978-0-471-30932-1
29. W.C. Chew, *Waves and Fields in Inhomogeneous Media*. IEEE PRESS Series on Electromagnetic Waves (IEEE Press, New York, 1995). ISBN 0-7803-4749-8

Chapter 2
The World of Plasmons

> *The world is full of magical things patiently waiting for our wits to grow sharper.*
>
> BERTRAND RUSSELL

Many of the fundamental electronic properties of the solid state can be described by the concept of single electrons moving between an ion lattice. If we ignore the lattice, in a first approximation, we end up with a different approach where the free electrons of a metal can be treated as an electron liquid of high density [1, 2]. From this plasma model it follows that longitudinal density fluctuations, so-called plasma oscillations or *Langmuir waves*, with an energy of the order of 10 eV will propagate through the volume of the metal. These volume excitations have been studied in detail with Electron Energy Loss Spectroscopy (EELS)[1] and have led to the discovery of surface plasmon polaritons.

2.1 From First Observations to the Modern Concept of Surface Plasmons

The first documented observation of surface plasmon polaritons dates back to 1902, when Wood[2] illuminated a metallic diffraction grating with polychromatic light and noticed narrow dark bands in the spectrum of the diffracted light, which he referred to as anomalies [4, 5].

Soon after Wood's measurements Lord Rayleigh[3] [6] suggested a physical interpretation of the phenomenon [7], but nevertheless it took several years until

---

[1] A historical overview of electron beam experiments to study surface plasmons can be found in [3], p. 47.

[2] Born 2nd May 1868 in Concord, Massachusetts; † 11th August 1955 in Amityville, New York.

[3] Born 12th November 1842 in Langford Grove, Essex; † 30th June 1919 in Witham, Essex. Nobel Prize in Physics 1904.

© Springer International Publishing Switzerland 2016
A. Trügler, *Optical Properties of Metallic Nanoparticles*, Springer Series in Materials Science 232, DOI 10.1007/978-3-319-25074-8_2

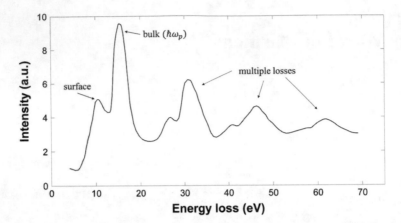

Fig. 2.1 First observation of surface plasmons, adapted with permission from [10], © 1959 by The American Physical Society. The surface plasmons were detected in the energy-loss spectrum of an Aluminum surface, where electrons with a kinetic energy of 2020 eV were specularly reflected under 45° incidence (also see [3])

Fano[4] [8] associated these anomalies with the excitation of electromagnetic surface waves on the diffraction grating. In 1957 Ritchie proposed the concept of surface plasmons in the context of electron energy loss in thin films [9] and the experimental verification followed 2 years later by Powell and Swan [10, 11], see Fig. 2.1.

In 1958 experiments with metal films on a substrate [12] again showed a large drop in optical reflectivity, and 10 years later the explanation and repeated optical excitation of surface plasmons were reported almost simultaneously by Otto [13] as well as Kretschmann and Raether [14]. They established a convenient method for the excitation of surface plasmons [4]. The principle of the plasmon excitation by Otto and Kretschmann is shown in Fig. 2.8 and discussed in the caption.

---

[4]Born 28th July 1912 in Turin; † 13th February 2001 in Chicago.

2.2 Derivation of Surface Plasmon Polaritons

Plasmons are (bosonic) elementary excitations in a metallic[5] solid. The basic ingredient for a material to allow such an excitation is the ability to form an intrinsic resonance, e.g. movable elementary charges that become displaced by an external field, experience a restoring force and start to oscillate. The question whether we should treat plasmons in a quantum mechanical or classical way is discussed in Sect. 3.1.

One of the most simple but nevertheless very utile models to describe the response of a metallic particle exposed to an electromagnetic field was proposed by Paul Drude[6] [20, 21] at the beginning of the twentieth century and further extended by Hendrik Lorentz[7] 5 years later (consult [22] for a detailed discussion). In 1933 Arnold Sommerfeld[8] and Hans Bethe[9] expanded the classical Lorentz-Drude model and eliminated some problems in the description of thermal electrons by accounting for the Pauli principle of quantum mechanics and replacing the Maxwell-Boltzmann with the Fermi-Dirac distribution (again see [22]).

Drude adopted a microscopic description of the electron dynamics in a metal in classical terms, and obtained the equation of motion of a damped oscillator where the electrons are moving between heavier, relatively immobile background ions:

Drude-Sommerfeld model of a free electron gas

$$m_e \frac{\partial^2 r}{\partial t^2} + m_e \gamma_d \frac{\partial r}{\partial t} = -eE_0 e^{-i\omega t}, \tag{2.1}$$

where γ_d describes a phenomenological damping term, m_e the effective free electron mass, e the free electron charge and ω and E_0 are the frequency and amplitude of the applied electric field respectively. Note that Eq. (2.1) describes the dynamic of one single electron, Drude just summed over many electrons later on to end up with a many-particle description. Equation (2.1) can be solved by the ansatz $r(t) = r_0 e^{-i\omega t}$ [23] which yields

$$r_0 = -\frac{\frac{e}{m_e}}{\omega^2 + i\gamma_d \omega} E_0. \tag{2.2}$$

---

[5]Recently it has been shown [15–19] that also doped graphene may serve as an unique two-dimensional plasmonic material with certain advantages compared to metals (lower losses and much longer plasmon lifetimes).

[6]Born 12th July 1863 in Braunschweig; † 5th July 1906 in Berlin.

[7]Born 18th July 1853 in Arnheim; † 4th February 1928 in Haarlem. Nobel Prize in Physics 1902.

[8]Born 5th December 1868 in Königsberg; † 26th April 1951 in München.

[9]Born 2th July 1906 in Straßburg; † 6th March 2005 in Ithaca, New York. Nobel Prize in Physics 1967.

If we now assume that the macroscopic polarization is given by a sum over charge times displacement r_0, we obtain

$$D = \varepsilon E = \varepsilon_0 E + P = \varepsilon_0 \left(1 - \frac{\omega_p^2}{\omega^2 + i\gamma_d \omega} \right) E, \qquad (2.3)$$

which finally leads to the dielectric function of Drude form

Dielectric function of Drude form for metals

$$\frac{\varepsilon_d(\omega)}{\varepsilon_0} = \varepsilon_\infty - \frac{\omega_p^2}{\omega^2 + i\gamma_d \omega}, \qquad \text{with} \qquad \omega_p = \sqrt{\frac{n_e e^2}{\varepsilon_0 m_e}}, \qquad (2.4)$$

where we have additionally accounted for the ionic background ε_∞ in a metal (additional screening introduced by the bound valence electrons of the positive ion cores). Here ω_p is the *volume* or *bulk plasma frequency* (electron density $n_e = 3/4\pi r_s^3$, r_s is the electron gas parameter or mean electron distance and takes the value 0.159 or 0.160 nm for gold or silver, respectively [22]), and ε_0 is the vacuum permittivity.[10] If we neglect γ_d and ε_∞ for the moment, the Drude dielectric function simplifies to $1 - \omega_p^2/\omega^2$ and we can distinguish two frequency regions: If ω is larger than ω_p, ε_d is positive and the corresponding refractive index $n = \sqrt{\varepsilon_d/\varepsilon_0}$ is a real quantity.[11] But if ω is smaller than ω_p, ε_d becomes negative and n is imaginary. An imaginary refractive index implies that an electromagnetic wave cannot propagate inside the medium. The specific value of $\hbar\omega_p$ for most metals lies in the ultraviolet region, which is the reason why they are shiny and glittering in the visible spectrum. A light wave with $\omega < \omega_p$ is reflected, because the electrons in the metal screen the light. On the other hand if $\omega > \omega_p$ the light wave gets transmitted (the metal becomes transparent), since the electrons in the metal are too slow and cannot respond fast enough to screen the field.

This treatment of a free electron gas already gives quite accurate results for the optical properties of metals in the infrared region, but since higher-energy photons can also promote bound electrons from lower-lying bands into the conduction band [23] (see Fig. 2.2) the Drude model becomes inaccurate for the visible regime as indicated in Fig. 2.3.

---

[10]For gold at room temperature we have $n_e = 3/[4\pi(0.159\times10^{-9})^3] = 5.9 \times 10^{28}$ electrons/m³. With $e = 1.602 \times 10^{-19}$ As, $\varepsilon_0 = 8.854 \times 10^{-12}$ A² s⁴/kg m³, and $m_e = 9.109 \times 10^{-31}$ kg we obtain $\omega_p = 13.75$ PHz or $\hbar\omega_p = 9.05$ eV, respectively.

[11]In general $n = \pm\sqrt{\mu\varepsilon/\mu_0\varepsilon_0} \approx \pm\sqrt{\varepsilon/\varepsilon_0}$ for optical frequencies and the positive sign is chosen for causality reasons in the system. A negative refractive index does not occur in nature but can be artificially generated with metamaterials (see Sect. 9 or e.g. [24–26]).

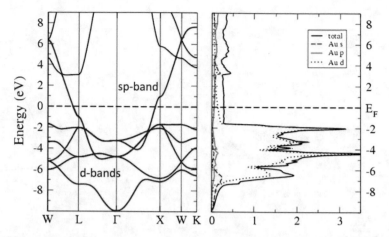

Fig. 2.2 Electronic band structure of gold calculated within a first-principles approach. The parabolic sp-bands (energy roughly proportional to momentum squared) explain why the free electron gas description works well for most metals. Above 2 eV (i.e. light wavelengths shorter than 620 nm) electrons can be promoted from the d-bands below the Fermi energy to states above, which leads to strong plasmon damping and the absorption as well as re-emission of light yielding the golden color. Adapted figure with permission from [27]. © 2004 by The American Physical Society

As will be discussed in Sect. 3.1 in more detail, the dielectric function describes the response of a material and can either be obtained by first principle calculations or from measurements. In our simulations we will employ the dielectric data[12] obtained from experiments, but unfortunately there are some difficulties. First the results from different experiments are not always consistent as shown in Fig. 2.4. In particular the data published in [30] shows some additional features compared to [28] for gold and silver around 1 eV ($\approx$1240 nm),[13] see real part in Fig. 2.4a and zig-zag features in Fig. 2.4b. Second the dielectric function can be determined from optical experiments on bulk solids, thin solid films or clusters [31] which all differ from each other. A more detailed comparison as well as spectroscopic ellipsometry measurements of evaporated, template-stripped, and single-crystal gold can be found in [32] for example.

---

[12]In the literature most of the time the complex refractive index is tabulated, the connection to the dielectric function is given by $\varepsilon/\varepsilon_0 = \varepsilon_1 + i\varepsilon_2 = n^2 = (\tilde{n} + i\tilde{k})^2$. The real and imaginary parts of $\varepsilon/\varepsilon_0$ then follow as $\varepsilon_1 = \tilde{n}^2 - \tilde{k}^2$ and $\varepsilon_2 = 2\tilde{n}\tilde{k}$.

[13]The conversion factor between eV and nm is 1239.84 as discussed in Appendix A.1.

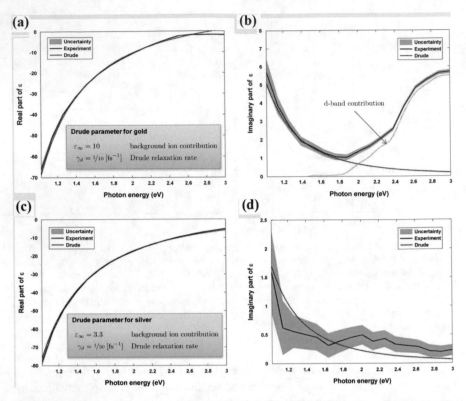

Fig. 2.3 Real and imaginary parts of the dielectric function for gold (**a**), (**b**) and silver (**c**), (**d**). The experimental data together with the measuring uncertainty have been taken from Johnson and Christy [28]. In the *inset* on the *left hand panels* we list the corresponding Drude parameters, see Eq. (2.4). The imaginary part of the Drude dielectric function for gold becomes invalid for energies above 1.9 eV (wavelengths below 650 nm), see (**b**), because at this energy interband transitions set in. The line for the d-band contribution in (**b**) is obtained from a simple comparison between the Drude dielectric function and the experimental result, also see [29]

2.2.1 Electromagnetic Waves at Interfaces

In Chap. 3 we will see that the wave equation (3.31) in Helmholtz form is the one relation to rule them all, electromagnetic fields always have to obey

$$\left(\nabla^2 + k^2 \right) E(r, \omega) = 0,$$

where the wave number k is given by $n\frac{\omega}{c}$. By following [23] let us investigate a planar interface between a metal and a dielectric with $\varepsilon_1 = \varepsilon_m$ for the metal at $z < 0$, and $\varepsilon_2 = \varepsilon_b$ for the dielectric at $z > 0$ (at optical frequencies we can set $\mu_{1,2} = \mu \approx \mu_0$). The wave equation now has to be solved separately in each region of constant ε and the corresponding boundary conditions demonstrate how to match

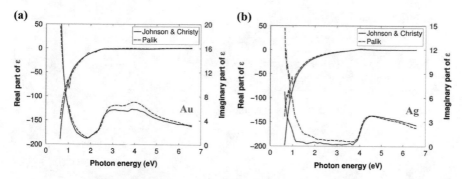

Fig. 2.4 Comparison of the dielectric data for gold (**a**) and silver (**b**) obtained from experiment and published in [28, 30], also see [32]

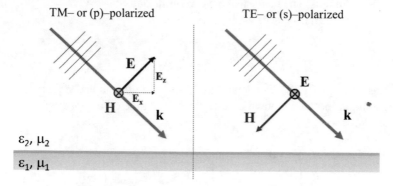

Fig. 2.5 Polarized electromagnetic waves at a planar interface

the two solutions at the interface. In general, Maxwell's equations allow two sets of self-consistent solutions with different polarizations–TM or (p)-polarized and TE or (s)-polarized modes,[14] see Fig. 2.5.

Since we do not obtain a plasmonic excitation for the latter (see [33] for example), we neglect TE modes and write down the solution of Eq. (3.31) as [23]:

$$E_j = \begin{pmatrix} E_{x_j} \\ 0 \\ E_{z_j} \end{pmatrix} e^{i(k_x x - \omega t)} e^{i k_{z_j} z}, \qquad j = 1, 2. \tag{2.5}$$

The component k_x of the wave vector parallel to the interface is preserved, thus the index j indicating the medium is unnecessary here. Applying the Pythagorean

---

[14]The abbreviations *(s)* and *(p)* come from the German words *senkrecht* (perpendicular) and *parallel* (parallel), respectively.

theorem to the wave vectors results in

$$k_x^2 + k_{z_j}^2 = k_j^2 \quad \longrightarrow \quad k_{z_j} = \sqrt{\frac{\varepsilon_j}{\varepsilon_0}\left(\frac{\omega}{c}\right)^2 - k_x^2}, \qquad j = 1, 2, \qquad (2.6)$$

and the boundary conditions (see Sect. 3.2.1) yield

$$E_{x_1} - E_{x_2} = 0, \qquad (2.7a)$$

$$\varepsilon_1 E_{z_1} - \varepsilon_2 E_{z_2} = 0, \qquad (2.7b)$$

i.e. the parallel field component is continuous, whereas the perpendicular component is discontinuous. The fields also have to fulfill Gauss' law $\nabla \cdot \mathbf{D} = 0$ [Eq. (3.3a)] in both source-free half-spaces which gives us

$$k_x E_{x_j} + k_{z_j} E_{z_j} = 0. \qquad (2.8)$$

With Eqs. (2.7) and (2.8) we now have a set of four coupled equations, which yield a solution for the unknown field components if the corresponding determinant vanishes. We get

$$\det \begin{pmatrix} 1 & 0 & -1 & 0 \\ 0 & \varepsilon_1 & 0 & -\varepsilon_2 \\ k_x & k_{z_1} & 0 & 0 \\ 0 & 0 & k_x & k_{z_2} \end{pmatrix} = k_x(\varepsilon_1 k_{z_2} - \varepsilon_2 k_{z_1}) \overset{!}{=} 0, \qquad (2.9)$$

i.e. for $k_x \neq 0$ we have $k_{z_1} = \frac{\varepsilon_1}{\varepsilon_2} k_{z_2}$. Together with Eq. (2.6) this directly yields the dispersion relation between the wave vector components and the angular frequency ω (also see [34])

Plasmon dispersion relation

$$k_x = \frac{\omega}{c}\left[\frac{\varepsilon_1 \varepsilon_2}{\varepsilon_0 (\varepsilon_1 + \varepsilon_2)}\right]^{1/2} = n_j \frac{\omega}{c}\left(\frac{\varepsilon_i}{\varepsilon_1 + \varepsilon_2}\right)^{1/2}, \qquad (2.10)$$

$$k_{z_j} = \frac{\omega}{c}\left[\frac{\varepsilon_j^2}{\varepsilon_0 (\varepsilon_1 + \varepsilon_2)}\right]^{1/2} = n_j \frac{\omega}{c}\left(\frac{\varepsilon_j}{\varepsilon_1 + \varepsilon_2}\right)^{1/2}, \qquad j = 1, 2, \ i = 2, 1.$$

$$(2.11)$$

We are looking for solutions that are propagating along the surface, i.e. we require a real k_x and a purely imaginary k_z. From the dispersion relation it follows that the

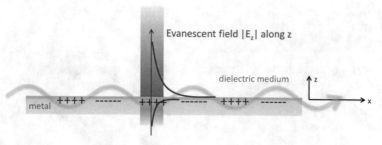

Fig. 2.6 An evanescent wave corresponds to a TM-mode that propagates along the interface of a metal and a dielectric, where the z-component of the electric field decays exponentially [35]

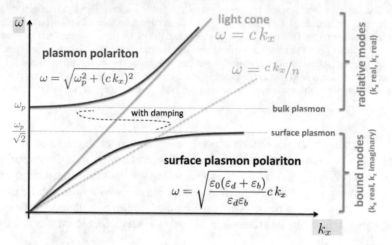

Fig. 2.7 Plasmon dispersion relation for a metal/air interface with $k_x(\omega) \in \mathbb{R}$. Since the dispersion line of plasmons (*red line*, without damping; *blue line* for free electrons) does not cross the light cone (*yellow line*) at any point, it is not possible to excite a surface plasmon at a metal air interface with a light wave. Yet the light cone can be tilted (*dotted yellow line*) if we change from free space to a dielectric medium. For $k_x \to \infty$ the denominator $\varepsilon_0(\varepsilon_1 + \varepsilon_2)$ in Eq. 2.10 should vanish, thus for $\varepsilon_1 = \varepsilon_d \approx \varepsilon_0(1 - \omega_p^2/\omega^2)$ and $\varepsilon_2 = \varepsilon_b = \varepsilon_0$ we derive the characteristic surface plasmon frequency $\omega_p/\sqrt{2}$.

conditions for the existence of an interface mode are given by

$$\varepsilon_1 \varepsilon_2 < 0, \qquad \varepsilon_1 + \varepsilon_2 < 0, \tag{2.12}$$

which results in an electromagnetic wave *bound to the interface*[15] as depicted in Fig. 2.6. The resulting imaginary k_{z_j} corresponds to exponentially decaying (so-called *evanescent*) waves.

---

[15]See [1] or [23] for a more detailed discussion.

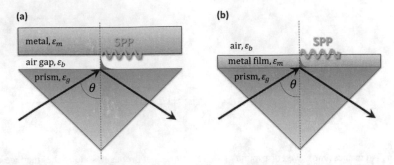

Fig. 2.8 Two configurations that provide the missing momentum contribution discussed in the text for the excitation of surface plasmons. (**a**) Otto configuration: The total reflection at the prism/air interface generates an evanescent field that excites a surface plasmon polariton at the dielectric/metal interface. This is an useful method when the metal surface should not be damaged, but it is difficult to keep the constant distance (usually of the order of λ) between the metal and the prism. (**b**) Kretschmann configuration: Here the total reflection at the prism/metal interface generates an evanescent field that excites a surface plasmon at the opposite metal/air interface. The thickness of the metal film must be smaller than the skin depth [see Eq. (2.13)] of the evanescent field

The dispersion relation (2.10) is plotted in Fig. 2.7. Since the red line of the surface plasmon polariton does not cross the light cone at any point, a direct excitation of surface plasmons with an electromagnetic wave is not possible (the momentum of light is always too small). Nevertheless surface plasmon polaritons can be excited, of course, one just has to provide the missing momentum contribution. This can be done through a tilt of the light line $\omega = ck_x$ to ck_x/n in a dielectric medium as shown in Figs. 2.7 and 2.8.

2.2.1.1 Skin Depth and Propagation Length

The imaginary part of the dielectric function is related to the energy dissipation of the material, i.e. if an electromagnetic wave impinges on a metal surface it can only penetrate the solid up to a certain material dependent depth. This so-called *skin depth* ζ is defined as the distance, where the exponentially decreasing evanescent field $e^{-|k_{z_j}||z|}$ falls to $1/e$ [1, 31]:

Skin depth

$$\zeta_j = \frac{1}{|k_{z_j}|} \quad \text{or} \quad \begin{cases} \zeta_b = \frac{\lambda}{2\pi\, n_b} \left[\frac{\varepsilon_m + \varepsilon_b}{\varepsilon_b} \right]^{1/2} & \text{in medium with } \varepsilon_b, \\[2mm] \zeta_m = \frac{\lambda}{2\pi\, n_m} \left[\frac{\varepsilon_m + \varepsilon_b}{\varepsilon_m} \right]^{1/2} & \text{in metal with } \varepsilon_m. \end{cases} \qquad (2.13)$$

Since the dielectric background constant ε_b is usually much smaller than the real part of ε_m, inside the metal (2.13) can be replaced with the approximation $\zeta_m \approx \lambda/(2\pi n'_m)$, where $n'_m = \sqrt{\varepsilon'_m/\varepsilon_0}$ and $\varepsilon_m = \varepsilon'_m + i\,\varepsilon''_m$.

If a surface plasmon propagates along a smooth surface, its intensity decreases as $e^{-2k''_x x}$ [1], where $k_x = k'_x + i\,k''_x$. The length $\delta = 1/2k''_x$ after which the intensity has fallen to $1/e$ can be defined as the *propagation length*.

In this sense the large real part of ε_m is responsible for the corresponding small skin depth.[16]

2.2.2 Particle Plasmons

In general, we have seen that plasmons arise from an interplay of electron density oscillations and the exciting electromagnetic fields. In this sense, we should talk about surface plasmon polaritons and also distinguish the propagating (evanescent) modes at the interface of a metal and a dielectric from their localized counterpart at the surface of metallic particles (so-called *particle plasmon polaritons*). If an electromagnetic wave impinges on a metallic nanoparticle (whose spatial dimension is assumed to be much smaller than the wavelength of light), the electron gas gets polarized (polarization charges at the surface) and the arising restoring force again forms a plasmonic oscillation, see Fig. 2.9.

The metallic particle thus acts like an oscillator and the corresponding resonance behavior determines the optical properties [36]. Such particle plasmons[17] behave

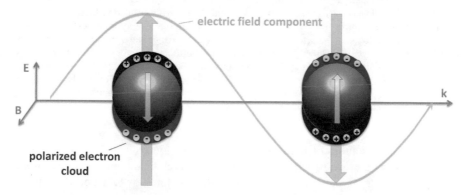

Fig. 2.9 Excitation of particle plasmons through the polarization of metallic nanoparticles. At the resonance frequency the plasmons are oscillating with a 90° phase difference (180° above resonance). In addition, a magnetic polarization occurs, but most of the time it can be neglected for reasons discussed in Sect. 2.8

---

[16]Typically ζ_m is one order of magnitude smaller than ζ_b, e.g. for gold we obtain $\zeta_m \approx 30$ nm and $\zeta_b \approx 280$ nm at $\lambda = 600$ nm.

[17]Since we solely discuss particle plasmon polaritons in this book, we will henceforth always refer to them.

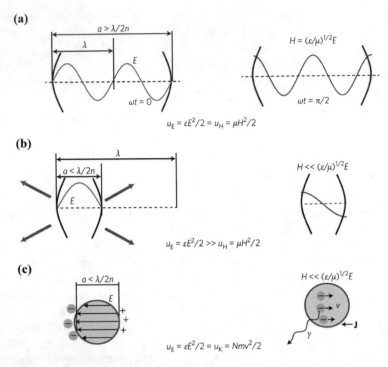

Fig. 2.10 Energy balance in photonic and plasmonic structures, reprinted by permission from Macmillan Publishers Ltd: Nature Nanotechnology [38], © 2015. (**a**) In an optical cavity with dimensions larger than half wavelength $\lambda/2n$ energy alternates between u_E (*left*), the energy of electric field E, and u_H (*right*), energy of magnetic field H, analogous to energy alternating between potential and kinetic forms for a mass oscillating on a spring (the energy is conserved). (**b**) In a subwavelength cavity with characteristic size a the magnetic energy is too small, hence the cavity radiates energy out (*purple arrows*) in agreement with the diffraction limit. (**c**) When free carriers are introduced a current J flows and the energy alternates between u_E and the kinetic energy of carriers u_K, where N, m and v are respectively the carrier density, mass and velocity. The diffraction limit is beaten, but the motion of carriers is strongly damped with the damping rate γ, and the surface plasmon polariton mode is lossy

as efficient nanosources of light, heat and energetic electrons [37] and provide a unique playground for a very multidisciplinary research field, ranging from biology through chemistry to optics, from communication technology to material sciences, from solar light harvesting to cancer therapy.

Recently Jacob Khurgin published an insightful review about the loss in plasmonic materials [38], in which he also presented an intuitive picture about the energy balance in plasmonic structures (reprinted in Fig. 2.10).

In an optical resonator with characteristic dimension a larger than the wavelength λ, see Fig. 2.10a, the electromagnetic energy is transferred every half period from the electric field energy $u_E \sim \frac{1}{2}\varepsilon E^2$ to the magnetic field energy $u_H \sim \frac{1}{2}\mu H^2$ and back, similar to the kinetic and potential energy of an oscillating mass on a spring [38]. If the size of the resonator becomes smaller than $\lambda/2n$ as shown in Fig. 2.10b, where n is the refractive index, it follows from Maxwell's equations that the magnetic energy is much smaller than the electric counterpart, which makes self-sustaining oscillations impossible and the cavity radiates energy out. If one introduces free carriers into the mode as plotted in panel (c), the energy can also be stored in the form of kinetic energy u_k of the carriers and the energy balance $u_H + u_k = u_E$ can be restored at certain frequencies of the surface plasmon. This kinetic oscillation of the carriers allows us to exceed the diffraction limit of light, as we will see later on in Chap. 5, but it also comes at a price, since the electrons in a metal get scattered very quickly and the cavity mode vanishes typically after about 10 fs.

2.3   Tuning the Plasmon Resonance

When a metallic nanoparticle is illuminated by white light, the plasmonic resonance
determines the color we observe, see Fig. 2.11. This behavior is nothing new:
Microscopic gold and silver particles incorporated in the stained glass of old church
windows are responsible for their beautiful lustrous colors.[18] Another very famous
example dates back to antiquity–a Roman cup made of dichroic glass illustrating
the myth of King Lycurgus[19] [41, 42].

Let us discuss this topic more precisely: We can tune the resonance of the surface
plasmon polariton by changing the size or shape of a metallic nanoparticle, as
plotted in Fig. 2.12. We recognize that the effect of squeezing a sphere to a rod-
like particle has a significantly greater impact than increasing its diameter and the
upper panel in Fig. 2.12b also shows that the resonance intensity has a maximum
for the aspect ratio of somewhere between 0.3 and 0.4 (also see [43]).

The plasmonic resonance is not only sensitive to the shape and size of a
nanostructure, also the dielectric medium surrounding the particle plays a key role,
see [44] for example. In Fig. 2.13 the sensitivity of a nanorod to the dielectric
background ε_b is shown.

Even the slightest change in the dielectric surrounding leads to a detectable shift
of the resonance energy. That is the reason why metallic nanoparticles are very

Fig. 2.11 Nanoparticles of various shape and size in solution–the plasmonic resonance determines
the color (Photo with kind permission from Carsten Sönnichsen, http://www.nano-bio-tech.de/)

---

[18]The windows of Sainte-Chapelle in Paris are a very good example of this: Light transmission
through the metal ions in the stained glass strongly depends on the incident and viewing angles. At
sunset, the grazing-angle scattering of light by gold particles in the window creates a pronounced
red glow that appears to slowly move downward, while intensities of blue tints from ions of copper
or cobalt remain the same [39, 40].

[19]Lycurgus cup, fourth century AD, British Museum, London.

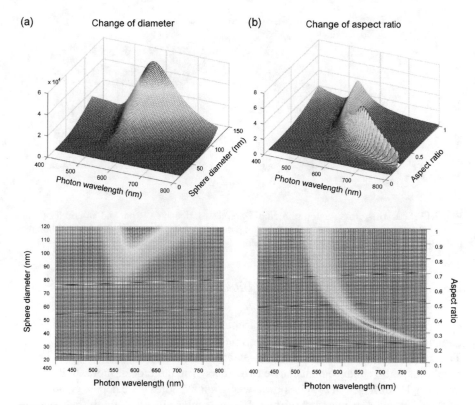

Fig. 2.12 Tuning the resonance of a surface plasmon polariton by changing the diameter of a gold nanosphere (**a**) or by squeezing its aspect ratio (**b**). The *upper panels* show the scattering cross section, in the *lower panels* the corresponding density plots are shown

suitable for sensing applications: Placing a molecule in the vicinity of a nanoparticle effects the dielectric environment and therefore shifts the plasmon peak. As already mentioned in the introduction, bio-sensing is one of the major applications for plasmonic nanoparticles. Thus the question naturally arises of the optimal shape and size of a sensor made of nanoparticles. A comprehensive analysis of this question can be found in [43, 45] for example.

The influence of the embedding medium on a metallic nanoparticle is a tricky topic as discussed in Sect. 4.6.3. For example, if particles in aqueous solution are investigated, a constant and homogeneous water temperature must be assured because the refractive index of water is temperature dependent [46]. Indeed, the change from 20 to 40 °C leads to a resonance shift of about 1 nm [47], which for very accurate sensing applications may become important–especially when heat is generated through an exciting laser field [48] (also see Sect. 2.9).

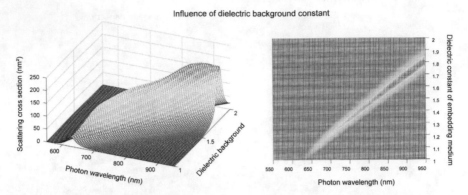

Fig. 2.13 Scattering cross section of a gold nanorod with diameter 10 nm and an arm length of 35 nm. The dielectric constant ε of the embedding medium varies from 1.0 to 2.0, the panel on the right again shows the corresponding density plot

2.3.1 Principle of Plasmonic (Bio-)sensing

One possible route to plasmonic molecular sensors is given by exploiting the enhancement of the decay rates of fluorophores in the vicinity of MNPs [49, 50]. The molecule uses the nanoparticle as an antenna [51] in order to emit its energy much faster–the enhancement can be two orders of magnitude or more. In Fig. 2.14 an example published in [49] is shown, where fluorophores were deposited onto two different samples of nanodisk-arrays. The molecules absorb in the ultraviolet and emit in the visible regime, which allows a separation of the excitation and emission channel.

If the plasmon frequency of the disk-arrays is in resonance with the molecular emission (blue dashed line in Fig. 2.14), each disk acts as a supplemental antenna for the molecules. The nonradiative near-field of the fluorophores gets converted into a radiating far-field, which leads to the dramatic increase of the radiative decay rate [49] as shown in Fig. 2.15.

The decay rate of the molecule can be calculated by Fermi's golden rule, as shown in [50]. In the cited work the decay rate of the coupled MNP-fluorophore system is derived, which is fully determined by the dyadic Green function of classical electrodynamics. It can be described in terms of a self-interaction, where the molecular dipole polarizes the MNP, and the total electric field acts, in turn, back on the dipole (see Fig. 2.16b).

It is not only the radiative decay rate which gets enhanced, the nonradiative decay channel is also strongly increased because of Ohmic power loss inside the nanodisks as indicated in Fig. 2.16 and discussed in [49].

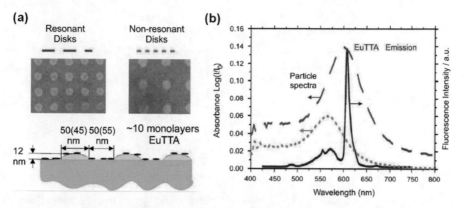

Fig. 2.14 (a) Electron-microscopic image of two different quadratic arrays of gold nanodisk samples on a silicon dioxide substrate, one geometry is chosen resonant and the other off-resonant to the fluorophore emission. The size parameters and mutual distances are shown in the *lower panel*. The large interparticle distance of ~50 nm ensures that coupling effects between the disks can be neglected. (**b**) Absorbance spectra (*dashed* and *dotted line*) of the two nanodisk samples as well as the emission spectrum (*solid line*, maximum at 612 nm) of the fluorophores. Reprinted figure with permission from [49]. © 2007 by The American Physical Society

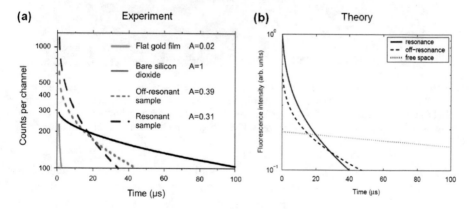

Fig. 2.15 Panel (**a**) shows the fitted fluorescence curves for the two samples of Fig. 2.14 (*dashed* and *dotted lines*), for a flat gold film (*yellow line* in *lower left corner*), and for the bare silicon substrate (*black solid line*). (**b**) Fluorescence intensity results of simulations with the boundary element method (see Sect. 4.4) for the corresponding setup, for further details see [49]. Reprinted figure with permission from [49]. © 2007 by The American Physical Society

This example illustrated the cumulative effect of many molecules spread over an array of nanoparticles, but the crux of the whole sensing problem is given by the possibility to detect one single molecule. Here once again the near-field enhancement of metallic nanoparticles is of great importance. In Fig. 2.17 the near-

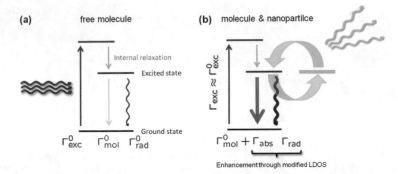

Fig. 2.16 Energy levels, radiative (Γ_{rad}) and non-radiative (Γ_{mol}) decay rates of a free molecule (**a**) and a molecule coupled to a metallic nanoparticle (**b**). The decay rates for the free fluorophore are indicated with a 0 in the superscript and the excitation is noted as Γ_{ext}. The plasmon resonance tuned to the molecular emission enhances both, the radiative as well as the nonradiative decay channel [49]

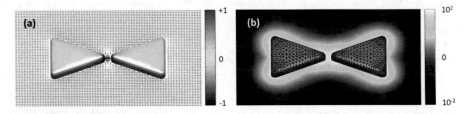

Fig. 2.17 (**a**) Real part of electric field (in the gap region the field is about seven times larger than the background) and normalized surface charge density, each one at the resonance energy for the gold bowtie nanoantenna with 5 nm gap distance. (**b**) Near-field enhancement $|E|^2/|E_0|^2$ calculated for the same nanoentenna. The triangle size is 45 nm in x, 40 nm in y and 8 nm in z direction

field enhancement of a bowtie nanoantenna is plotted on a logarithmic scale. For a gap distance of 5 nm we obtain a strong intensity enhancement as well as a localization in the gap region. Placing a single molecule in the hot spot at the gap leads one way to single molecule sensors, see Fig. 7.4. In [52], for example, single molecule fluorescence enhancements up to a factor of 1340 for gold bowtie nanoantennas have been reported.

Additional remark

An interesting review about advances in the field of optical-biosensors can be found in [53]. A typical example of the working principle of a sensor based on

Additional remark

surface plasmons is shown in Fig. 2.18 below, where the binding of analytes can be measured non-invasively in real time.

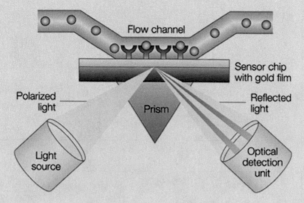

Fig. 2.18 The changes in the refractive index in the immediate vicinity of a surface layer are detected with a sensor chip. The plasmonic resonance is observed as a sharp shadow in the reflected light at an angle that depends on the mass of material at the surface–this angle shifts if biomolecules bind to the surface. Reprinted by permission from Macmillan Publishers Ltd: Nature Reviews Drug Discovery [53], © 2002

2.3.2 Surface-Enhanced Raman Scattering

Another important technique that allows the detection of single molecules is surface-enhanced Raman scattering (SERS) [54, 55]. Through the absorption of a photon, a molecule can be excited electronically and also vibrationally. For typical fluorescence the received energy is spontaneously re-emitted after some internal relaxation (see Fig. 2.16a). In quite rare events also elastic or inelastic scattering may occur, where the photon frequency can be mixed with the vibrational energy levels of the molecule, see Fig. 2.19. The elastic process, where the photon is absorbed and emitted with the same frequency, is called *Rayleigh scattering*. If the electronic transition also involves the vibrational energy levels of the molecule, we end up with the inelastic *Raman scattering*.[20]

---

[20]A unified treatment of fluorescence and Raman scattering can be found in [56] for example.

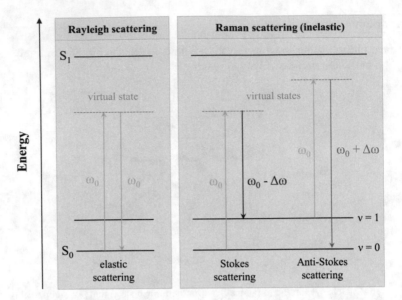

Fig. 2.19 Jablonski diagram of (elastic) Rayleigh and (inelastic) Raman scattering. The electronic levels $S_{0,1}$ (sometimes also called bands) of a molecule may each possess a number of vibrational sublevels $v = 0, 1, \ldots$. If the energy difference $\Delta\omega$ is subtracted from or added to the original photon energy ω_0, we obtain Stokes or Anti-Stokes scattering, respectively

Raman scattering can also simply be understood as the analog of amplitude modulation used in broadcasting [23]: The frequency of the carrier (laser) is mixed with the frequencies of the signal (molecular vibrations). Hence the final Raman signal consists of sums and differences of these frequencies and is a highly specific fingerprint of the investigated molecule. The problem in this process is the very low probability for a photon to undergo Raman scattering. A typical Raman cross section is up to 14 orders of magnitude smaller than the fluorescence cross section [23], which makes it very difficult to investigate Raman scattering in microscopy. However, in the late 1970s an enormous increase of the Raman signal for molecules adsorbed on specially prepared silver surfaces was reported [57–59] and led to the birth of a new research field [60]. The Raman signal, increased at least a millionfold, results from the interaction with the strong plasmonic near fields at the surface of rough metallic films or nanoparticles. The reason for the vast SERS signal (e.g. in contrast to the previously discussed fluorescence enhancement) lies in a twofold amplification[21]: First, the incident light field which excites the Raman modes is magnified through the plasmonic excitation and then the Raman signal itself is further enhanced a second time through the same process [55]. Each

---

[21]Also an enhancement in polarizability due to chemical effects such as charge-transfer excited states can contribute to the huge SERS signal [55], but in the following we will just stick to the electromagnetic enhancement.

enhancement step is proportional to E^2, which finally yields a E^4 field dependence for the SERS signal. Accordingly the enhancement factor E_S can be expressed as

Electromagnetic SERS enhancement

$$E_S \approx |E(\omega)|^2 |E(\omega')|^2 \approx |E(\omega)|^4, \qquad (2.14)$$

where ω represents the incident frequency and ω' the corresponding Stokes-shifted value. Usually the plasmon width is large compared to the Stokes shift and allows the approximation $E_S \propto |E|^4(\omega)$.[22] A more rigorous expression for E_S has been developed in [62], but since numerical values differ only slightly from (2.14), the simpler expression is almost exclusively used [55]. In conventional SERS, the enhancement is averaged over the surface area of the particle where molecules can adsorb to generate the observed enhancement factor $\langle E_S \rangle$, while in single-molecule SERS it is the maximum enhancement that is of interest [55]. Nowadays SERS enhancement factors up to 10^{11} are regularly reported [63] thereby paving the way for the detection of single molecules [64, 65].

----

[22]Some studies on isolated, homogeneous particles have shown that this assumption leads to a slight overestimate of the enhancement factor [55]. Also see [61] for example.

2.4   The Energy Transfer of Förster and Dexter

Light-sensitive molecules play a decisive role in our everyday life, be it in photosynthesis and plant growth or the marvelous diversity of colors and dyes surrounding us every minute. Such *color bearers* are called chromophores, their color arises when the molecule absorbs photons at specific optical wavelengths and transmits, reflects or re-emits (fluorescence) the absorbed electromagnetic energy. The involved electronic transitions in the molecular orbitals take place in the visible spectrum of light and thus are detectable to the human eye.

If two chromophores come close to each other, an initially excited donor can transfer its energy to an acceptor through nonradiative processes, which are essential for photosynthesis for example (see [66] for a more detailed discussion). These energy transfers may occur intermolecularly, i.e. between two different molecules, or intramolecularly between two different parts of the same molecule, provided that the emission spectrum of the donor overlaps with the absorption spectrum of the acceptor, see Fig. 2.20.

This naturally results in fluorescence quenching for the donor, i.e. the fluorescence intensity is decreased, whereas the acceptor's intensity is increased. If the mutual spatial separation of donor and acceptor is smaller than approximately 5 nm, the nonradiative energy transfer can occur through a classical dipole-dipole interaction, provided that the molecular dipole moments of both partners are oriented in similar directions (usually there is a statistical distribution of dipole moments). If the separation between donor and acceptor becomes smaller than approximately 0.5 nm, the electronic orbitals start to overlap and the electrons can be transferred bilaterally between donor and acceptor through a quantum mechanical process, see Fig. 2.21.

The first mechanism is called Förster resonance energy transfer (FRET) [67, 68] and its efficiency is proportional to the inverse sixth power of the distance between

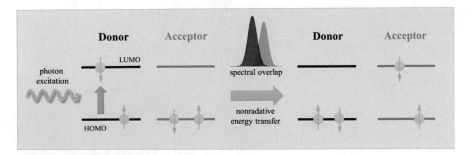

Fig. 2.20 Nonradiative energy transfer between a donor (*red*) and an acceptor (*green*). The donor initially absorbs a photon and gets excited. If an acceptor in its ground state is in close proximity to the donor and if the emission of the donor spectrally overlaps with the absorption of the acceptor, the energy can be transmitted *without* emission and absorption of an actual photon

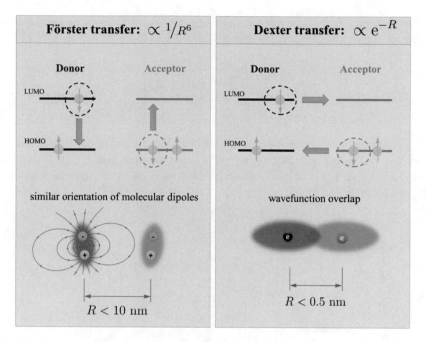

Fig. 2.21 Schematics of the singlet-singlet Förster and Dexter energy transfer mechanisms between an initially excited donor (*green*) and an acceptor (*red*). The Dexter energy transfer can also occur for a triplet state

donor and acceptor. The second mechanism is called Dexter transfer [69] and follows an inverse exponential law.

The FRET efficiency is usually defined as the relative change of the donor's fluorescence emission [23]

$$E_{\text{FRET}} = \frac{\gamma^{D \to A}}{\gamma^{D} + \gamma^{D \to A}}, \tag{2.15}$$

where γ^{D} is the sum of the radiative and molecular decay rate of the donor in absence of the acceptor and $\gamma^{D \to A}$ is the rate of energy transfer from donor to acceptor, which also depends on the particular dipole orientations. Usually these orientations are unknown and a statistical average over many donor-acceptor pairs is required to determine $\gamma^{D \to A}$. The corresponding equations are derived in detail in [23], for example, but here we will just present the final result for the averaged transfer rate

$$\frac{\gamma^{D \to A}}{\gamma^{D}} = \left(\frac{R_{\text{F}}}{R}\right)^{6}, \tag{2.16}$$

where $R = |r_D - r_A|$ is the distance between donor and acceptor and the *Förster radius* R_F is given by Novotny and Hecht [23]

$$R_F = \left[\frac{3c^4}{4\pi} \int_0^\infty \frac{f_D(\omega)\sigma_A(\omega)}{n(\omega)^4\omega^4} \, d\omega \right]^{-1/6}. \tag{2.17}$$

Here f_D is the donor's normalized emission spectrum in a medium with refractive index[23] n, σ_A is the acceptor's absorption cross section, and c is the speed of light. Thus Eq. (2.17) is basically given by the spectral overlap between donor and acceptor for averaged dipole orientations. The FRET rate (2.15) then follows as

FRET rate for averaged dipole orientations

$$E_{FRET} = \frac{\gamma^{D \to A}}{\gamma^D + \gamma^{D \to A}} = \frac{1}{1 + \left(\frac{R}{R_F} \right)^6}. \tag{2.18}$$

Typical values for the Förster radius R_F range between 2 and 9 nm [71]. For short distances ($kR \ll 1$) the transfer rate $\gamma^{D \to A}$ always scales as R^{-6}, for long distances ($kR \gg 1$) and aligned dipoles we obtain a $(kR)^{-4}$ dependence, and if the dipoles are not aligned, $\gamma^{D \to A}$ decays as $(kR)^{-2}$ [23].

Because of the antenna-like behavior of metallic nanoparticles [51], it doesn't come as a surprise that they can be used as mediators for the Förster energy transfer between separated molecules. The donor then excites a plasmon and the plasmon in turn passes on its energy to the acceptor through the FRET mechanism, i.e. through dipole-dipole coupling. In [72], for example, the authors used an europium[3+] complex as donor, which emits a narrow line at 612 nm when excited by light at 360 ±20 nm. As acceptor they used a Cy5 dye, which absorbs in a broad band mainly between 580 and 680 nm. The Förster radius for this donor-acceptor pair yields 5.6 nm and a result for the FRET probability is shown in Fig. 2.22.

---

[23]The usage of the refractive index in Eq. (2.17) is not consistent in the literature, see [70] for more details.

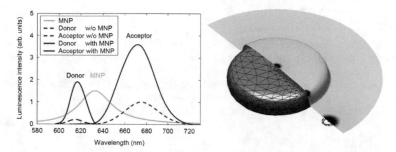

Fig. 2.22 Simulated FRET probability where a MNP is used as mediator between donor and acceptor molecules. On the left fluorescence spectra with (*solid lines*) and without (*dashed lines*) the metallic nanoparticle are shown. The *bright, orange line* indicates the scattered intensity of the metallic nanoparticle in absence of donor and acceptor molecules. On the right the surface discretization of the cylinder-shaped particle (60 nm diameter and 15 nm height) as used in the calculations is depicted. The donor and acceptor molecules are placed on a sheet with a distance 2 nm away from the MNP. The *color map* shows the FRET probability for three selected donor positions. The *solid* and *dashed lines* indicate the effective Förster radii for the donor-acceptor complex in the presence and absence of the MNP. Reprinted with permission from [72]. © 2008 American Chemical Society

---

Additional remark

The coupling between metallic nanoparticles and dyes or other quantum emitters can be exploited in several ways. In 2003, David Bergman and Mark Stockman proposed a coherent and ultrafast source of optical energy concentrated at the nanoscale and named it in analogy to its light-amplifying brother *spaser* (short for surface plasmon amplification by stimulated emission of radiation) [73], see Fig. 2.23.

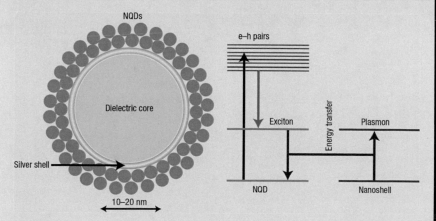

Fig. 2.23 Schematic of a spaser and the corresponding energy levels and transitions. The spaser is made from a silver nanoshell on a dielectric core (with a radius of 10–20 nm), and

Additional remark

surrounded by two dense monolayers of nanocrystal quantum dots. The external radiation excites a transition into electron-hole (e-h) pairs (*vertical black arrow* in the *right panel*). The e-h pairs relax to excitonic levels (*green arrow*). The exciton recombines and its energy is transferred (without radiation) to the plasmon excitation of the metal nanoparticle (nanoshell) through resonant coupled transitions (*red arrows*). Reprinted by permission from Macmillan Publishers Ltd: Nature Photonics [74], © 2008

In [74] Mark Stockman writes:

A spaser is the nanoplasmonic counterpart of a laser, but it (ideally) does not emit photons. It is analogous to the conventional laser, but in a spaser photons are replaced by surface plasmons and the resonant cavity is replaced by a nanoparticle, which supports the plasmonic modes. Similarly to a laser, the energy source for the spasing mechanism is an active (gain) medium that is excited externally.

The reason that surface plasmons in a spaser can work analogously to photons in a laser is that their relevant physical properties are the same. First, surface plasmons are bosons: they are vector excitations and have spin 1, just as photons do. Second, surface plasmons are electrically neutral excitations. And third, surface plasmons are the most collective material oscillations known in nature, which implies they are the most harmonic (that is, they interact very weakly with one another). As such, surface plasmons can undergo stimulated emission, accumulating in a single mode in large numbers, which is the physical foundation of both the laser and the spaser.

Through the combination with metamaterials the concept was further developed to so-called *lasing spasers* in [75], also a *graphene spaser* was proposed recently [76]. Attempts to realize spasers in the laboratory usually have to face the strong absorption losses in metals particularly at optical frequencies. A first spaser-based nanolaser, consisting of 44-nm-diameter nanoparticles with a gold core and a dye-doped silica shell, was reported in [77].

2.5   Light Absorption in Solar Cells

Another of the many possible applications that could be revolutionized (or at least very much improved) by plasmonic structures are photovoltaic devices [78, 79]. The combination of MNPs and semiconductor materials, for example, allows a considerable reduction in the physical thickness of absorber layers, and yields new possibilities for the design of solar-cells. We don't want to go into the details of this broad research field, but since some of the solar cell aspects are again prime examples for the tunability of the plasmonic resonance, let us briefly dwell upon this subject. A review about recent advances at the intersection of plasmonics and photovoltaics can be found in [79] and a brief overview of three different light-trapping thin-film geometries is depicted in Fig. 2.24.

Unfortunately for the developing of photovoltaic cells our sun is a black body source that has nothing to do with the compliant light of a laser. An efficient solar device must for example have a very broad absorption spectrum for all parts of sun light (also see [80]). There are two easy ways to broaden the absorption of a metallic nanoparticle, either by adding an additional coating around the structure or simply by coupling it to another particle, see Fig. 2.25.

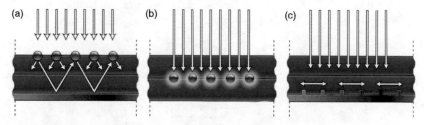

Fig. 2.24 Three different plasmonic geometries for light-trapping for thin-film solar cells. (**a**) Metallic nanoparticles embedded at the solar cell surface scatter light preferentially into the semiconductor thin film which leads to an increase of the optical path length in the cell. (**b**) If the particles are embedded in the semiconductor, the creation of electron-hole pairs is caused by the particle's near-field. (**c**) Light coupling through a corrugated metal back surface. Reprinted by permission from Macmillan Publishers Ltd: Nature Materials [79], © 2010

38   2 The World of Plasmons

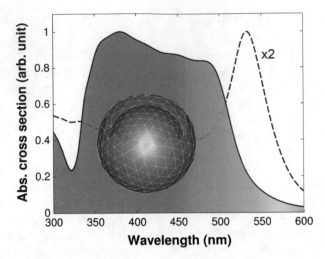

Fig. 2.25 The *blue dashed line* shows the absorption spectrum for a 10 nm gold sphere embedded in glass ($n = 1.5$). Adding an additional silver layer around the sphere (see *inset*), yields an enhanced and broadened absorption (*red solid line*, normalized to 1). The *shaded background color* range was approximated according to Fig. A.1

Additional remark

The coupling of two or more metallic nanoparticles leads to a hybridization of the energy levels. Figure 2.26 shows the example for spherical dimers.

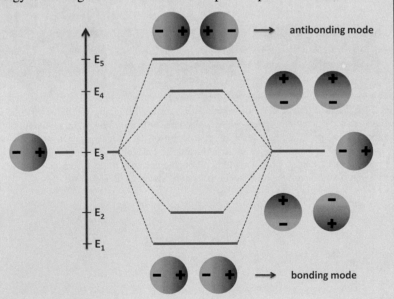

Fig. 2.26 Energy levels of two coupled spherical nanoparticles (see [81, 82]), note the occurrence of bonding and antibonding modes

2.6 Strong Coupling

In the last sections the interaction of molecules and metallic nanoparticles took
center stage several times. Possible applications of such hybrid systems cover a
wide area and recently also quantum-optics found its way into plasmonics [83, 84],
where the key element is given by the strong coupling of a quantum emitter and a
resonator [85]. This allows for a coherent transfer of the excitation energy between
emitter and resonator, which was first observed for single atoms in high-finesse optical resonators [86, 87]. More recently, strong coupling has been reported for various
solid state systems, such as semiconductor quantum dots [88, 89] or superconductor
circuits [90]. If a metallic nanoparticle interacts with a molecule we usually obtain
an intricate interplay of the molecule-MNP coupling strength with the molecular
relaxation dynamics, which becomes heavily altered in the vicinity of the nanoparticle. This typically results in excitonic splitting (*Rabi splitting*[24]), frequency shifts,
asymmetric line shapes (*Fano resonances*[25]), or dips in the scattering spectra [94].

If we assume a generic model, where a quantum emitter couples only to a single
cavity mode (e.g. the dipole mode), we can write down a simple expression for the
onset of strong coupling in terms of the coupling strength g [85]

Strong coupling regime

$$g > \frac{1}{4} |\gamma_c - \gamma_m|,\tag{2.19}$$

where γ_c and γ_m are the decay rates of the cavity and quantum emitter, respectively.
In simple terms: We obtain strong coupling if the coupling strength is larger than
the damping in the system. In plasmonic systems a priori this scenario is doubtful,
since we always have to deal with strong losses in the metals. On the other hand
the coupling strength in hybrid structures is also very high (e.g. see [95]) and strong
coupling in plasmonic systems has been observed in several experiments to date.
But in general, Eq. (2.19) is too simple, because if we take a metallic nanoparticle
as cavity and a molecule as emitter, the coupling is not restricted only to the
nanoparticle dipole mode but can also occur to all other modes. Nevertheless (2.19)
allows an estimation of the pertinent parameters for strong coupling. For example

---

[24]The formation of two hybridized modes (cf. bonding and antibonding modes in Fig. 2.26)
oscillating at different energies is a typical indicator of strong coupling, if the involved damping is
small. In the time domain the energy then oscillates between atom and cavity (Rabi oscillations,
named after the Nobel laureate Isidor Isaac Rabi) [91]. In the frequency domain we obtain Rabi
splitting and anticrossing, see [88]. Rabi splitting is fundamental for the dynamics of two-state
systems and can be easily modeled using the Jaynes-Cummings model [92] for example.

[25]Fano interference occurs, when a resonant or discrete state interacts with a continuum of states–
a very general effect that can be found in many different areas of physics and has been derived
originally by Ugo Fano [93].

for a silver nanoparticle the plasmon decay rate is roughly $30\,\mathrm{fs}$, so the critical coupling strength follows as $g \approx {}^{30}/_4 \sim 5\,\mathrm{meV}$ [85]. For gold we have $\gamma_c = 10\,\mathrm{fs}$ and therefore $g \approx 1.7\,\mathrm{meV}$.

A unified theoretical framework of strong coupling between plasmons and electronic transitions is discussed in [94]. There the authors show that by modifying the damping rate of a plasmon resonance (e.g. by changing size, shape, or nature of the metal), it is possible to transition from one regime of coupling to another (e.g., from Rabi splitting to Fano interference), see Figs. 2.27 and 2.28. A quantum mechanical approach can be found in [96] for example, also see [97–99].

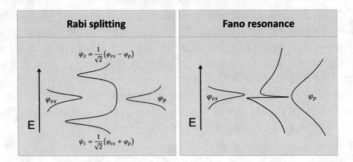

Fig. 2.27 Illustration of Rabi splitting and a Fano line shape in a hybrid plasmonic/excitonic system. On the *left panel* a plasmonic oscillator with a sufficiently narrow line width couples strongly with an excitonic resonance to produce two hybridized modes split in energy. In the *right panel* an oscillator with an extremely broad line width resembles a continuum of states and undergoes Fano interference with the excitonic resonance. Adapted with permission from [94]. © 2014 American Chemical Society

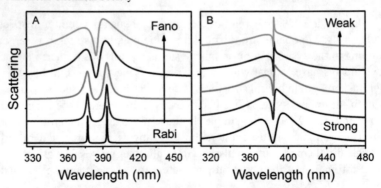

Fig. 2.28 Generalized model of plasmonic/excitonic coupling predicting various distinct regimes of coupling, see [94] for details. (**a**) A coupled oscillator model shows that as the damping rate or line width of the plasmonic resonator is increased, there is a transition from a Rabi splitting regime to a Fano interference regime. Intermediate cases are also shown to demonstrate the gradual evolution between the two regimes. (**b**) Decreasing the value of the coupling constant results in a transition from an antiresonance feature in the Lorentzian spectrum to an asymmetric line shape superimposed on the Lorentzian spectrum. Intermediate cases demonstrate the gradual transition between these two Fano regimes. Figure and caption reprinted with permission from [94]. © 2014 American Chemical Society

2.7 Damping Mechanisms of Surface Plasmons

As discussed at the beginning of this chapter, a plasmon is formed when a coherent charge density oscillation is induced in the electron gas of a metallic nanoparticle by an external excitation. This collective motion of the electrons can easily be disturbed, for example by scattering events that destroy the phase coherence. One can imagine this dephasing by simply kicking an electron out of the lock-step march, due to scattering with impurities, phonons, other electrons, and so on. The electron still has its kinetic energy but the phase coherence gets destroyed. Figure 2.29 gives an overview about the different decay channels and in Fig. 2.30 the corresponding time scales are plotted.

The radiative and nonradiative break up and decay processes of plasmons result in highly excited electron-hole pairs [38, 100], which thermalize by further collision processes on a sub-ps time scale [101, 102] to a distribution of "hot" electrons[26] and holes [36]. Finally through electron-phonon coupling the energy is transferred to the lattice as heat. Typically after about 10 fs the plasmon oscillation decays and the further relaxation can last from femtoseconds up to several nanoseconds, see Fig. 2.30.

The decay time τ of a particle plasmon oscillation can be determined from the homogeneous linewidth Γ of the spectral resonance of a plasmon, see Appendix A.2. Γ is defined as the full width at half maximum (FWHM) and is inversely proportional to the decay time τ. $\Gamma \propto \tau^{-1}$, see Eq. (A.7). Figure 2.31 shows an example for a nanorod antenna.

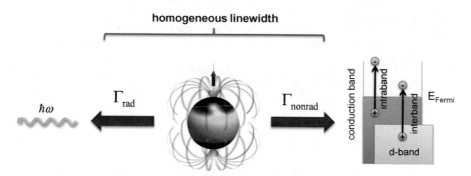

Fig. 2.29 Usually a particle plasmon shares the destiny of a mayfly, albeit on a different time scale: After a quite short existence it is doomed to decay. It can either decay radiatively (*left*) and emit photons or lose its energy non radiatively via intra- and interband transitions (*right*), also see Fig. 2.30. Both decay channels contribute to the homogeneous linewidth Γ

---

[26] Since the heat capacity of the electronic system is much smaller than that of the ion lattice, an excitation by femtosecond laser pulses can generate extremely high electron temperatures [36].

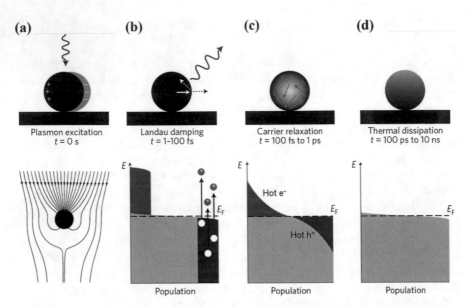

(a)   **(b)**   **(c)**   **(d)**

Plasmon excitation
$t = 0$ s

Landau damping
$t = 1$–100 fs

Carrier relaxation
$t = 100$ fs to 1 ps

Thermal dissipation
$t = 100$ ps to 10 ns

Fig. 2.30 Photoexcitation and subsequent relaxation processes following the illumination of a metal nanoparticle with a laser pulse, and characteristic timescales. Reprinted by permission from Macmillan Publishers Ltd: Nature Nanotechnology [100], © 2015. (**a**) First, the excitation of a localized surface plasmon redirects the flow of light (Poynting vector) toward and into the nanoparticle. (**b–d**) Schematic representations of the population of the electronic states (*gray*) following plasmon excitation: hot electrons are represented by the *red areas* above the Fermi energy E_F and hot hole distributions are represented by the *blue area* below E_F. (**b**) In the first 1–100 fs Landau damping occurs, where the athermal distribution of electron-hole pairs decays either through re-emission of photons or through electron-electron interactions. During this very short time interval, the hot carrier distribution is highly non-thermal. (**c**) The hot carriers will redistribute their energy by electron-electron scattering processes on a timescale ranging from 100 fs to 1 ps. (**d**) Finally, heat is transferred to the surroundings of the metallic structure on a longer timescale ranging from 100 ps to 10 ns, via thermal conduction

Increased damping, for example caused by defects in the nanoparticles' crystal structure, thus leads to a broadening of the spectral linewidth. Note that Γ can vary more than a factor of ten for different nanoparticle geometries–defining the quality of a plasmonic sensor simply over the shift of the resonance may therefore be a little bit simplistic. In [43] we introduce several 'figures of merit' to allow a better quality comparison of different plasmonic sensors, also see [45]. As discussed in Chap. 7, for the direct measurement of the temporal evolution of particle plasmons ultrashort laser pulses are necessary. These pulses have become available in the past decade, enabling the observation of ultrafast plasmon dynamics directly in the time domain with fs time-resolution [103].

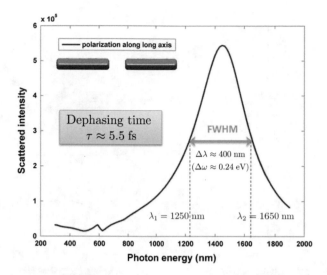

Fig. 2.31 Homogeneous linewidth (FWHM) Γ and plasmon decay time τ for a gold nanorod antenna (rod length 280 nm, width 60 nm, height 40 nm, and gap distance 65 nm). The resulting decay time $\tau \approx 5.5$ fs for this geometry has also been verified by autocorrelation measurements

In general, high absorption loss is usually the crux of the matter in plasmonics. A recent summary of the four main absorption processes in a metal gives a clear depiction [38], see the simplified band structure scheme in Fig. 2.32. The two states below and above the Fermi level correspond to an excited electron with wave vector k_2 and energy $E_2 = \hbar\omega$, and the remaining hole with wavevector k_1 and energy E_1. The magnitude difference between these two momenta is typically far too large to be supplied by a photon ($\Delta k_{1,2} \approx 3$ nm^{-1}) [38], hence in Fig. 2.32a the missing momentum part comes from a phonon or imperfection, where we obtain a hot electron and hole. The other momentum-conserving absorption process is electron-electron scattering [panel (b)], which is strongly frequency dependent and where the energy conservation relation $E_3 + E_4 = E_1 + E_2 + \hbar\omega$ indicates that we end up with four 'lukewarm' carriers with kinetic energies of the order of $\hbar\omega/4$ each. The next process in panel (c) is Landau damping, where the absorption is associated with the finite size of the surface plasmon mode and leads to an additional confinement contribution (also see Chap. 8). In panel (d) again the discussed band-to-band absorption is shown, a mechanism inherent to all metals.[27]

---

[27]At the beginning of this section we have seen that d-band absorption for gold starts around 620 nm, in silver around 400 nm.

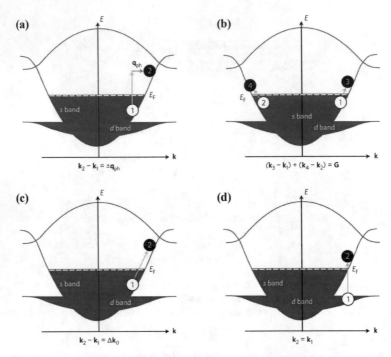

Fig. 2.32 Absorption of a quantum of electromagnetic energy $\hbar\omega$ in a metal, reprinted by permission from Macmillan Publishers Ltd: Nature Nanotechnology [38], © 2015. (**a**) Absorption assisted by a phonon with wavevector $\boldsymbol{q}_{\mathrm{ph}}$ creates hot hole (1) with wavevector $\boldsymbol{k}_1$ and hot electron (2) with wavevector $\boldsymbol{k}_2$, with energy E_2 above the Fermi level, E_F. (**b**) Electron-electron scattering assisted absorption, an *Umklapp process* involving reciprocal lattice vector $\boldsymbol{G}$ creates four lukewarm carriers–two holes (1,2) and two electrons (3,4). (**c**) Direct absorption (Landau damping) assisted by the plasmon momentum Δk_0 creates hot hole (1) and hot electron (2). (**d**) Interband absorption from the d to s shell does not create hot carriers because hole (1) in the d band has low velocity and excited electron (2) is too close to the Fermi level

2.8 Magnetic Effects

A light wave always consists of an oscillating electric and magnetic field, the one
never occurring without the other. But since at optical frequencies the value of the
magnetic permeability $\mu = \mu_0(1 + \chi_m) \approx \mu_0$, the magnetic component of light
generally plays an insignificant role and can often be neglected. This aspect can
be easily understood in a simplified picture with the Lorentz force F_L [104–107].
As has been highlighted above, the electron cloud of a metallic nanoparticle can
interact with an impinging light wave. In classical electrodynamics the effect of the
electromagnetic field on a moving charge q is described through [106, 107]

Lorentz force

$$F_L = q\left[E + (v \times B)\right].\tag{2.20}$$

The magnitude of the electric force is given by qE, the magnetic equivalent can be
expressed through qvB. In an electromagnetic wave we have $B/E \approx 1/c$ [108] (the
fraction is exactly the inverse of c in vacuum, in a dielectric we additionally would
have to account for the corresponding refractive index) and thus the ratio of the
velocity $|v|$ of the moving charge to the speed of light c determines the ratio of the
magnetic contribution to its electric counterpart. This ratio is essentially given by the
fine-structure constant α of atomic physics[28] [105]. For atoms we obtain $\alpha \approx 1/137$,
whereas in solid state physics the norm of the charge velocity is roughly given by
the Fermi velocity[29] v_F which implies for the ratio that $v_F/c \approx 1/300$ [105]. The
magnetic response of a material is determined by the magnetic susceptibility χ_m,
which scales as $(v_F/c)^2$. Two important conclusions follow from that: The magnetic
response is four orders of magnitude weaker than the ease with which the same
material is polarized [104] and magnetization in non-ferromagnetic materials is a
relativistic effect (also see Sect. 9.2)!

When we try to detect light in experiments, we are most of the time blind to
its magnetic part and can only perceive its electric component [104]. One way
to visualize both the magnetic- and electric-field distribution of propagating light

---

[28]This fundamental constant can be interpreted in several ways, e.g. as the electromagnetic
coupling strength for the interaction between electrons and photons, as a ratio of charges, energies,
or characteristic lengths. When Arnold Sommerfeld introduced this dimensionless number in 1916
to explain the splitting or *fine structure* of the energy levels of the hydrogen atom, which had been
observed, he considered the ratio of the velocity of the electron in the first circular orbit of the Bohr
model to the speed of light in vacuum.

[29]Numerical value for gold and silver particles $v_F \approx 1.4$ nm/fs $= 1400$ km/s, Fermi energy E_F
≈ 5.53 eV respectively [22]. In contrast, the typical *drift velocity* of electrons in an electric wire is
of the order of mm/hour.

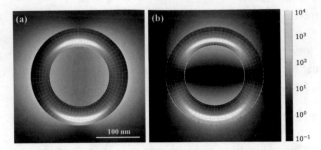

Fig. 2.33 (**a**) Logarithmic plot of the electric field intensity of a gold nanoring at the resonance frequency. The corresponding surface charge shows a dipolar distribution for a plane wave excitation polarized along the x-axis. The loop current for this particle design enhances the magnetic field (see Sect. 9.2) plotted in (**b**), but it is still approximately five times weaker than the electric field. The particle surface in (**b**) also shows the magnetic field

has been demonstrated by Burresi et al. in [104] through the combination of near-field measurements and metamaterials. Sometimes the special geometry of a nanoparticle also induces a magnetic field, e.g. ring-shaped particles sustaining loop currents (see Fig. 2.33). Recently attempts have been made to intertwine magnetism and plasmonics to so-called *magnetoplasmonics*, where usually noble metals are combined with ferromagnetic materials to tailor the magneto-optical properties of nanostructures, see the reviews [109] or [110] for example.

2.9 Temperature Dependence and Coupling to Lattice Vibrations

The Boltzmann constant k_B (another bridge between macroscopic and microscopic physics) has the dimension of energy divided by temperature and thus allows to connect temperature with a thermodynamic energy value. In this sense we are able to assign an energy of 0.025 eV to the typical room temperature–a value much smaller than the plasmonic energies at optical frequencies (1.65–3.26 eV, see Fig. A.1). Consequently in plasmonic experiments an explicit temperature dependence can usually be neglected, as long as the excitation does not melt away the metal of course.[30]

It has been mentioned above that the refractive index of a medium surrounding a metallic nanoparticle may be temperature dependent and that, e.g. for water, a reasonable temperature change of 20 °C already leads to a detectable resonance shift [46, 47]. Because of the extreme sensitivity of surface plasmons to their direct surrounding we can exploit such thermal induced changes once again for sensor applications or very local sensitivity measurements. The latter becomes possible through the utilization of thermosensitive polymers like PNIPAM [112, 113], see Fig. 2.34. Such stimulable plasmonic systems are very efficient candidates in *active plasmonics*,[31] since they provide a continuous and reversible modulation of the plasmonic response [112]. Also the subsequent decay of electron-hole pairs into phonons finally leads to the decomposition of plasmons into lattice heat. However, the typical Debye temperature for metals is of the order of the room temperature [115], thus electron-phonon coupling usually plays an insignificant role compared to plasmonic interactions. Nevertheless, an increase of the intrinsic electron-lattice interactions in metal clusters with sizes smaller than 10 nm has been reported in [116] and similar as for magnetic effects, the expansion to materials other than plain gold or silver again allows us to overcome certain restraints. Hence the coupling of plasmons and phonons has been reported for semiconductors and graphene [117], for example. Phonon-enhanced light matter interaction at the nanoscale with surface phonon polaritons as infrared counterparts to surface plasmons has been introduced in [118]. One of the advantages of this approach is the weaker damping for phonons, which thus allows for stronger and sharper optical resonances.

---

[30]Optical damage of metal nanoparticles usually starts at laser energies around 25 GW/cm² for antenna structures. For single particles this intensity might be doubled because of the lower field enhancement, see e.g. [111].

[31]In real-life applications some kind of active control over the properties of the corresponding nanosystem is usually required to achieve signal switching, modulation or amplification, for example. For a passive device these properties are fixed by the nanostructure parameters, in active plasmonics typically hybridized systems (see e.g. [114]) are used, where metallic nanostructures are combined with functional materials.

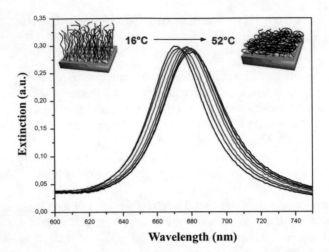

Fig. 2.34 Principle of an active polymer-coated plasmonic nanostructure. The PNIPAM polymer undergoes a reversible phase transition upon going from a hydrophilic swollen conformational state to a hydrophobic collapsed state around its lower critical solution temperature at 32 °C in pure water. A set of eight extinction spectra of the PNIPAM-coated nanoparticles arrays going from 16 to 52 °C are shown. Reprinted with permission from [112]. © 2011 American Chemical Society

Another case where temperature effects become important will be briefly discussed in Sect. 6.3, where thermal heating is used to reduce surface roughness of metallic nanoparticles. Such tempering processes usually also change the crystallite grain sizes of the metal and lead to a modified dielectric function.

Recently a set of refractory materials also gained interest amongst the plasmonic community: Transition metal nitrides [119, 120] such as titanium nitride (TiN) or zirconium nitride (ZrN) mimic the optical properties of gold and silver but are chemically stable at temperatures above 2000 °C.

2.10   Nanoparticle Fabrication Methods

The huge increase of research activities about plasmonic nanoparticles in the last decades was made possible as a result of the improvement of nanofabrication methods and the involved advancement of the control of matter at the nanoscale. Two main techniques are widely used for the fabrication of metallic nanoparticles: Chemically synthesis and electron beam lithography [121] together with lift-off based vapor deposition. A discussion of these methods as well as further references can be found in [122] for example.

2.10.1   Chemical Synthesis

A widely-used seeded-growth technique for chemical synthetisation of gold nanoparticles was proposed by Nikobaakht et al. in [123], also see [124]. With their approach, rod-shaped particles are grown in a two step method [125]. First a gold salt is quickly reduced in an aqueous medium to elementary gold in metallic form, resulting in nanospheres with an added organic molecule as a shell to prevent their aggregation. During the second part a small amount of the previous dispersion is added to a solution containing gold salt in a slower reductive medium, promoting the reaction to metallic gold in the surface of the gold nanospheres. With the organic molecule shell presenting the tendency to bind better to specific crystal orientations, the growth of the particles in a specific direction is favored, leading to rod-like, single-crystalline structures. Lengths for a nanorod synthesized through this route vary between a few and 50 nm, with 25–30 nm breadth and relatively small size deviations.

Nanoparticles made of silver can be obtained by reducing silver nitrate with ethylene glycol in the presence of a water-soluble polymer and sodium sulfide as described in [47, 127]. The particle shape can be controlled by the temperature and the reaction time (with the original method from [127] silver nanocubes are fabricated, applying a higher temperature and a longer reaction time gives rod-shaped particles [47]), see Fig. 2.35. Silver triangles for example can be synthesized by a photo-induced process, where spherical silver colloids are transformed into triangular nanoplates [47, 128].

2.10.2   Electron Beam Lithography

In the following Fig. 2.36 the principle process steps of nanoparticle fabrication with electron beam lithography are shown, see [129] or [121] for example.

1. One starts with a cleaned substrate surface, whose material is chosen accordingly to the desired application. If the material is non-conductive (like a glass

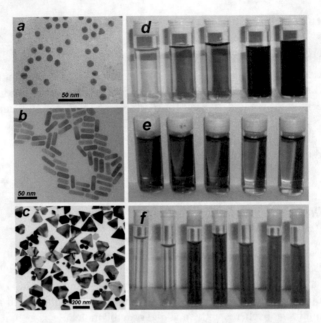

Fig. 2.35 Examples of synthesized nanoparticles, reprinted from [126], © 2004, with permission from Elsevier. Left: Transmission electron micrographs of Au nanospheres and nanorods (**a, b**) and Ag nanoprisms (**c**) (mostly truncated triangles) formed using citrate reduction, seeded growth, and DMF reduction, respectively. Right: Photographs of colloidal dispersions of AuAg alloy nanoparticles with increasing Au concentration (**d**), Au nanorods of increasing aspect ratio (**e**), and Ag nanoprisms with increasing lateral size (**f**)

substrate), a conductive layer of indium-tin oxide (ITO) or metal has to be deposited on the substrate (high vacuum vapor deposition).

2. In a spin-coat process a typically 100 nm thick layer of PMMA is put on the structure. This layer is a positive electron resist and has to be baked on a heating plate at 170 °C for 8 h [129].

3. The PMMA layer is exposed by an electron beam and the desired nanoparticle structure is written.

4. After the electron beam exposure the resist layer is chemically developed. (Starting with a developer bath for 30 s, immediately followed by a chemical stopper bath for another 30 s and an isopropanol rinsing step [129].)

5. After cleaning, the nanoparticle metal (e.g. gold) is deposited in a high vacuum evaporation process.

6. An acetone bath is used to remove the remaining PMMA layer resist.

7. To finish the fabrication process, the surface is cleaned with a final isopropanol rinsing.

The resulting particles are polycrystalline and their surface roughness is higher than that of chemical synthesized particles (also see Chap. 6). Thermal annealing

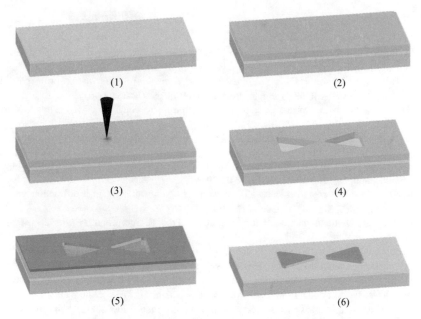

Fig. 2.36 Fabrication steps of electron beam lithography as described in the text. (*1*) Cleaned substrate surface with ITO, (*2*) spin coated PMMA layer, (*3*) writing nanostructure with e-beam, (*4*) chemical development, (*5*) high vacuum metal evaporation (*6*) lift off and cleaning

could help to reduce the roughness and leads to reduced damping, an increasing absorption and the FWHM becomes narrower, see [125].

Another similar technique is *ion beam lithography*, where much heavier charged particles are used instead of a beam of electrons.

References

1. H. Raether, *Surface Plasmons on Smooth and Rough Surfaces and on Gratings*. Springer Tracts in Modern Physics, vol. 111 (Springer, Berlin, 1988). ISBN 978-0387173634
2. H. Raether, *Excitation of Plasmons and Interband Transitions by Electrons*. Springer Tracts in Modern Physics, vol. 88 (Springer, Berlin, 1980). ISBN 978-3540096771
3. F.J. García de Abajo, Optical excitations in electron microscopy. Rev. Mod. Phys. **82**, 209 (2010).
4. J. Dostálek, J. Homola, S. Jiang, J. Ladd, S. Löfås, A. McWhirter, D.G. Myszka, I. Navratilova, M. Piliarik, J. Štěpánek, A. Taylor, H. Vaisocherová, *Surface Plasmon Resonance Based Sensors*. Springer Series on Chemical Sensors and Biosensors (Springer, Berlin/Heidelberg, 2006). ISBN 978-3642070464
5. R.W. Wood, On a remarkable case of uneven distribution of light in a diffraction grating spectrum. Philos. Mag. **4**(21), 396–402 (1902).
6. L. Rayleigh, On the dynamical theory of gratings. Proc. R. Soc. Lond. A **79**(532), 399–416 (1907).
7. R.B. Schasfoort, A.J. Tudos (eds.), *Handbook of Surface Plasmon Resonance* (Royal Society of Chemistry, Cambridge, 2008). ISBN 978-0854042678
8. U. Fano, The theory of anomalous diffraction gratings and of quasi-stationary waves on metallic surfaces (Sommerfeld's waves). J. Opt. Soc. Am. **31**, 213–222 (1941).
9. R. Ritchie, Plasma losses by fast electrons in thin films. Phys. Rev. **106**(5), 874 (1957).
10. C.J. Powell, J.B. Swan, Origin of the characteristic electron energy losses in aluminum. Phys. Rev. **115**, 869 (1959).
11. C.J. Powell, J.B. Swan, Effect of oxidation on the characteristic loss spectra of aluminum and magnesium. Phys. Rev. **118**, 640–643 (1960).
12. T. Turbadar, Complete absorption of light by thin metal films. Proc. Phys. Soc. **73**(1), 40 (1959).
13. A. Otto, Excitation of nonradiative surface plasma waves in silver by the method of frustrated total reflection. Zeits. Phys. **216**(4), 398–410 (1968).
14. E. Kretschmann, H. Raether, Radiative decay of nonradiative surface plasmons excited by light. Z. Naturforsch. A **23**, 2135–2136 (1968)
15. F.H.K. Koppens, D.E. Chang, F.J. García de Abajo, Graphene plasmonics: a platform for strong light-matter interaction. Nano Lett. **11**, 3370–3377 (2011).
16. A.N. Grigorenko, M. Polini, K.S. Novoselov, Graphene plasmonics. Nat. Photonics **6**(11), 749–758 (2012).
17. A. Vakil, N. Engheta, Transformation optics using graphene. Science **332**(6035), 1291–1294 (2011).
18. F.J. García de Abajo, Graphene nanophotonics. Science **339**(6122), 917–918 (2013).
19. F.J. García de Abajo, Graphene plasmonics: challenges and opportunities. ACS Photon. **1**(3), 135–152 (2014).
20. P. Drude, Zur Elektronentheorie der Metalle. Ann. Phys. **306**(3), 566–613 (1900).
21. P. Drude, Zur Elektronentheorie der Metalle; II. Teil. Galvanomagnetische und thermomagnetische Effecte. Ann. Phys. **3**, 369–402 (1900).
22. N.W. Ashcroft, N.D. Mermin, Festkörperphysik. Oldenbourg, München, 2007. ISBN 978-3-48658273-4
23. L. Novotny, B. Hecht, *Principles of Nano-Optics*, 2nd edn. (Cambridge University Press, Cambridge, 2012). ISBN 978-1107005464
24. J.B. Pendry, Negative refraction makes a perfect lens. Phys. Rev. Lett. **85**, 3966–3969 (2000).

25. H.J. Lezec, J.A. Dionne, H.A. Atwater, Negative refraction at visible frequencies. Science **316**, 5823 (2007).
26. N.I. Zheludev, A roadmap for metamaterials. Opt. Photonics News **22**, 30–35 (2011).
27. F. Ladstädter, U. Hohenester, P. Puschnig, C. Ambrosch-Draxl, First-principles calculation of hot-electron scatterings in metals. Phys. Rev. B **70**, 235125 (2004).
28. P.B. Johnson, R.W. Christy, Optical constants of the noble metals. Phys. Rev. B **6**, 12 (1972).
29. C. Sönnichsen, *Plasmons in metal nanostructures*. Ph.D. thesis, Fakultät für Physik der Ludwig-Maximilians-Universität München, 2001.
30. E.D. Palik (ed.), *Handbook of Optical Constants of Solids* (Academic Press, New York, 1985). ISBN 978-0125444200
31. U. Kreibig, M. Vollmer, *Optical Properties of Metal Clusters*. Springer Series in Material Science, vol. 25 (Springer, Berlin, 1995). ISBN 978-3-540-57836-9
32. R. Olmon, B. Slovick, T. Johnson, D. Shelton, S.-H. Oh, G. Boreman, M. Raschke, Optical dielectric function of gold. Phys. Rev. B **86**, 235147 (2012).
33. S.A. Maier, *Plasmonics: Fundamentals and Applications* (Springer, Berlin, 2007). ISBN 978-0-387-33150-8
34. F.-P. Schmidt, H. Ditlbacher, U. Hohenester, A. Hohenau, F. Hofer, J.R. Krenn, Universal dispersion of surface plasmons in flat nanostructures. Nat. Commun. **5**, 3604 (2014).
35. W.L. Barnes, A. Dereux, T.W. Ebbesen, Surface plasmon subwavelength optics. Nature **424**(6950), 824–830 (2003).
36. B. Lamprecht, *Ultrafast plasmon dynamics in metal nanoparticles*. Ph.D. thesis, Institut für Physik, Karl-Franzens-Universität Graz, 2000.
37. G. Baffou, R. Quidant, Nanoplasmonics for chemistry. Chem. Soc. Rev. **43**, 3898–3907 (2014).
38. J.B. Khurgin, How to deal with the loss in plasmonics and metamaterials. Nat. Nano. **10**(1), 2–6 (2015).
39. M.I. Stockman, Nanoplasmonics: the physics behind the applications. Phys. Today **64**, 39–44 (2011).
40. M.I. Stockman, Dark-hot resonances. Nature **467**, 541–542 (2010).
41. U. Leonhardt, Optical metamaterials: invisibility cup. Nat. Photonics **1**, 207–208 (2007).
42. H. Tait (ed.), *Five Thousand Years of Glass*, revised edition (University of Pennsylvania Press, Pennsylvania, 2004). ISBN 978-0-8122-1888-6
43. J. Becker, A. Trügler, A. Jakab, U. Hohenester, C. Sönnichsen, The optimal aspect ratio of gold nanorods for plasmonic bio-sensing. Plasmonics **5**(2), 161–167 (2010).
44. J. Pérez-Juste, I. Pastoriza-Santos, L.M. Liz-Marzán, P. Mulvaney, Gold nanorods: synthesis, characterization and applications. Coord. Chem. Rev. **249**, 1870–1901 (2005).
45. A. Jakab, C. Rosman, Y. Khalavka, J. Becker, A. Trügler, U. Hohenester, C. Sönnichsen, Highly sensitive plasmonic silver nanorods. ACS Nano **5**(9), 6880–6885 (2011).
46. S.K. Mitra, N. Dass, N.C. Varshneya, Temperature dependence of the refractive index of water. J. Chem. Phys. **57**, 1798 (1972).
47. J. Becker, *Plasmons as Sensors*. Ph.D. thesis, Fachbereich Chemie, Pharmazie und Geowissenschaften der Johannes-Gutenberg Universität Mainz, 2010.
48. A. Ohlinger, S. Nedev, A.A. Lutich, F.J., Optothermal escape of plasmonically coupled silver nanoparticles from a three-dimensional optical trap. Nano Lett. **11**, 1770 (2010).
49. S. Gerber, F. Reil, U. Hohenester, T. Schlagenhaufen, J.R. Krenn, A. Leitner, Tailoring light emission properties of fluorophores by coupling to resonance–tuned metallic nanostructures. Phys. Rev. B **75**, 073404 (2007).
50. U. Hohenester, A. Trügler, Interaction of single molecules with metallic nanoparticles. IEEE J. Sel. Top. Quantum Electron. **14**, 1430 (2008).
51. L. Novotny, N. van Hulst, Antennas for light. Nat. Photonics **5**(2), 83–90 (2011).

52. A. Kinkhabwala, Z. Yu, S. Fan, Y. Avlasevich, K. Müllen, W.E. Moerner, Large single-molecule fluorescence enhancements produced by a bowtie nanoantenna. Nat. Photonics 3, 654–657 (2009).
53. M.A. Cooper, Optical biosensors in drug discovery. Nat. Rev. Drug Discov. 1, 515–528 (2002).
54. P. Johansson, H. Xu, M. Käll, Surface-enhanced Raman scattering and fluorescence near metal nanoparticles. Phys. Rev. B 72, 035427 (2005).
55. K. Kneipp, M. Moskovits, H. Kneipp (eds.), *Surface-Enhanced Raman Scattering*. Topics in Applied Physics, vol. 103 (Springer, Heidelberg/New York, 2006). ISBN 978-3-540-33566-5
56. H. Xu, X.-H. Wang, M.P. Persson, H.Q. Xu, M. Käll, P. Johansson, Unified treatment of fluorescence and raman scattering processes near metal surfaces. Phys. Rev. Lett. 93, 243002 (2004).
57. M. Fleischmann, P. Hendra, A. McQuillan, Raman spectra of pyridine adsorbed at a silver electrode. Chem. Phys. Lett. 26(2), 163–166 (1974).
58. D.L. Jeanmaire, R.P. Van Duyne, Surface raman spectroelectrochemistry: Part I. Heterocyclic, aromatic, and aliphatic amines adsorbed on the anodized silver electrode. J. Electroanal. Chem. Interfacial Electrochem. 84(1), 1–20 (1977).
59. M.G. Albrecht, J.A. Creighton, Anomalously intense Raman spectra of pyridine at a silver electrode. J. Am. Chem. Soc. 99(15), 5215–5217 (1977).
60. M. Moskovits, Surface-enhanced spectroscopy. Rev. Mod. Phys. 57, 783 (1985).
61. E.C. Le Ru, J. Grand, N. Félidj, J. Aubard, G. Lévi, A. Hohenau, J.R. Krenn, E. Blackie, P.G. Etchegoin, Experimental verification of the SERS electromagnetic model beyond the | E | ^{4} approximation: polarization effects. J. Phys. Chem. C 112, 8117–8121 (2008).
62. M. Kerker, Estimation of surface-enhanced raman scattering from surface-averaged electromagnetic intensities. J. Colloid Interf. Sci. 118(2), 417–421 (1987).
63. E.C. Le Ru, E. Blackie, M. Meyer, P.G. Etchegoin, Surface enhanced raman scattering enhancement factors: a comprehensive study. J. Phys. Chem. C 111(37), 13794–13803 (2007).
64. S. Nie, S.R. Emory, Probing single molecules and single nanoparticles by surface enhanced Raman scattering. Science 275, 1102 (1997).
65. K. Kneipp, Y. Wang, H. Kneipp, L.T. Perelman, I. Itzkan, R.R. Dasari, M.S. Feld, Single molecule detection using surface-enhanced Raman scattering (SERS). Phys. Rev. Lett. 78, 1667 (1997).
66. W. Hoppe, W. Lohmann, H. Markl, H. Ziegler (eds.), *Biophysics* (Springer, Berlin/New York, 1983). ISBN 978-3-642-61816-1
67. T. Förster, Zwischenmolekulare Energiewanderung und Fluoreszenz. Ann. Phys. 437(1–2), 55–75 (1948).
68. R. Roy, S. Hohng, T. Ha, A practical guide to single-molecule FRET. Nat. Meth. 5(6), 507–516 (2008).
69. D.L. Dexter, A theory of sensitized luminescence in solids. J. Chem. Phys. 21(5), 836 (1953).
70. R.S. Knox, H. van Amerongen, Refractive index dependence of the Förster resonance excitation transfer rate. J. Phys. Chem. B 106(20), 5289–5293 (2002).
71. P. Wu, L. Brand, Resonance energy transfer: methods and applications. Anal. Biochem. 218(1), 1–13 (1994).
72. F. Reil, U. Hohenester, J.R. Krenn, A. Leitner, Förster-type resonant energy transfer influenced by metal nanoparticles. Nano Lett. 8(12), 4128–4133 (2008).
73. D. Bergman, M. Stockman, Surface plasmon amplification by stimulated emission of radiation: quantum generation of coherent surface plasmons in nanosystems. Phys. Rev. Lett. 90(2), 027402 (2003).

74. M.I. Stockman, Spasers explained. Nat. Photonics **2**(6), 327–329 (2008).
75. N.I. Zheludev, S.L. Prosvirnin, N. Papasimakis, V.A. Fedotov, Lasing spaser. Nat. Photonics **2**(6), 351–354 (2008).
76. V. Apalkov, M.I. Stockman, Proposed graphene nanospaser. Light Sci. Appl. **3**, e191 (2014).
77. M.A. Noginov, G. Zhu, A.M. Belgrave, R. Bakker, V.M. Shalaev, E.E. Narimanov, S. Stout, E. Herz, T. Suteewong, U. Wiesner, Demonstration of a spaser-based nanolaser. Nature **460**(7259), 1110–1112 (2009).
78. V.E. Ferry, L.A. Sweatlock, D. Pacifici, H.A. Atwater, Plasmonic nanostructure design for efficient light coupling into solar cells. Nano Lett. **8**, 4391 (2008).
79. H.A. Atwater, A. Polman, Plasmonics for improved photovoltaic devices. Nat. Mat. **9**, 205–213 (2010).
80. S. Divitt, L. Novotny, Spatial coherence of sunlight and its implications for light management in photovoltaics. Optica **2**(2), 95–103 (2015).
81. V. Myroshnychenko, J. Rodríguez-Fernández, I. Pastoriza-Santos, A.M. Funston, C. Novo, P. Mulvaney, L.M. Liz-Marzán, F.J.G. de Abajo, Modelling the optical response of gold nanoparticles. Chem. Soc. Rev. **37**, 1792–1805 (2008).
82. J.-S. Huang, J. Kern, P. Geisler, P. Weinmann, M. Kamp, A. Forchel, P. Biagioni, B. Hecht, Mode imaging and selection in strongly coupled nanoantennas. Nano Lett. **10**, 2105 (2010).
83. D.E. Chang, A.S. Sorensen, P.R. Hemmer, M.D. Lukin, Quantum optics with surface plasmons. Phys. Rev. Lett. **97**, 053002 (2006).
84. I.A. Akimov, J.T. Andrews, F. Henneberger, Stimulated emission from the biexciton in a single self-assembled II–VI quantum dot. Phys. Rev. Lett. **96**, 067401 (2006).
85. A. Trügler, U. Hohenester, Strong coupling between a metallic nanoparticle and a single molecule. Phys. Rev. B **77**, 115403 (2008).
86. Q.A. Turchette, C.J. Hood, W. Lange, H. Mabuchi, H.J. Kimble, Measurement of conditional phase shifts for quantum logic. Phys. Rev. Lett. **75**, 4710 (1995).
87. J.M. Raimond, M. Brunce, S. Haroche, Manipulating quantum entanglement with atoms and photons in a cavity. Rev. Mod. Phys **73**, 565 (2001).
88. K. Hennessy, A. Badolato, M. Winger, D. Gerace, M. Atatüre, S. Gulde, S. Fält, E.L. Hu, A. Imamoglu, Quantum nature of a strongly coupled single quantum dot-cavity system. Nature **445**, 896 (2007).
89. D. Press, S. Götzinger, S. Reitzenstein, C. Hofmann, A. Löffler, M. Kamp, A. Forchel, Y. Yamamoto, Photon antibunching from a single quantum-dot-microcavity system in the strong coupling regime. Phys. Rev. Lett. **98**, 117402 (2007).
90. A. Wallraff, D.I. Schuster, A. Blais, L. Frunzio, R.S. Huang, J. Majer, S. Kumar, S.M. Girvin, R.J. Schoelkopf, Strong coupling of a single photon to a superconducting qubit using circuit quantum electrodynamics. Nature **431**, 162 (2004).
91. G. Khitrova, H.M. Gibbs, M. Kira, S.W. Koch, A. Scherer, Vacuum Rabi splitting in semiconductors. Nat. Phys. **2**(2), 81–90 (2006).
92. S.M. Barnett, P.M. Radmore, *Methods in Theoretical Quantum Optics* (Clarendon, Oxford, 1997). ISBN 0-19-856362-0
93. U. Fano, Effects of configuration interaction on intensities and phase shifts. Phys. Rev. **124**, 1866–1878 (1961).
94. J.A. Faucheaux, J. Fu, P.K. Jain, Unified theoretical framework for realizing diverse regimes of strong coupling between plasmons and electronic transitions. J. Phys. Chem. C **118**(5), 2710–2717 (2014).
95. P. Anger, P. Bharadwaj, L. Novotny, Enhancement and quenching of single–molecule fluorescence. Phys. Rev. Lett. **96**, 113002 (2006).

96. A. Manjavacas, F.J. García de Abajo, P. Nordlander, Quantum plexcitonics: strongly interacting plasmons and excitons. Nano Lett. **11**(6), 2318–2323 (2011).
97. B. Gallinet, O.J.F. Martin, *Ab initio* theory of Fano resonances in plasmonic nanostructures and metamaterials. Phys. Rev. B **83**, 235427 (2011). doi:10.1103/PhysRevB.83.235427. http://journals.aps.org/prb/pdf/10.1103/PhysRevB.83.235427
98. B. Gallinet, O.J.F. Martin, Influence of electromagnetic interactions on the line shape of plasmonic Fano resonances. ACS Nano **5**, 8999–9008 (2011). doi:10.1021/nn203173r. http://pubs.acs.org/doi/pdf/10.1021/nn203173r
99. A. Lovera, B. Gallinet, P. Nordlander, O.J.F. Martin, Mechanisms of Fano resonances in coupled plasmonic systems. ACS Nano **7**, 4527–4536 (2013). doi:10.1021/nn401175j. http://pubs.acs.org/doi/pdf/10.1021/nn401175j
100. M.L. Brongersma, N.J. Halas, P. Nordlander, Plasmon-induced hot carrier science and technology. Nat. Nanotechnol. **10**(1), 25–34 (2015).
101. C.-K. Sun, F. Vallée, L.H. Acioli, E.P. Ippen, J.G. Fujimoto, Femtosecond-tunable measurement of electron thermalization in gold. Phys. Rev. B **50**, 15337–15348 (1994).
102. M. Wolf, Femtosecond dynamics of electronic excitations at metal surfaces. Surf. Scie. **377–379**, 343–349 (1997).
103. T. Hanke, J. Cesar, V. Knittel, A. Trügler, U. Hohenester, A. Leitenstorfer, R. Bratschitsch, Tailoring spatiotemporal light confinement in single plasmonic nanoantennas. Nano Lett. **12**(2), 992–996 (2012).
104. M. Burresi, D. van Oosten, T. Kampfrath, H. Schoenmaker, R. Heideman, A. Leinse, L. Kuipers, Probing the magnetic field of light at optical frequencies. Science **326**, 550–553 (2009).
105. H. Giessen, R. Vogelgesang, Glimpsing the weak magnetic field of light. Science **23**, 529–530 (2009).
106. J.D. Jackson, *Classical Electrodynamics* (Wiley, New York, 1962). ISBN 978-0-471-30932-1
107. A. Zangwill, *Modern Electrodynamics* (Cambridge University Press, Cambridge, 2012). ISBN 9780521896979
108. M. Pelton, G.W. Bryant, *Introduction to Metal-Nanoparticle Plasmonics* (Wiley/Science Wise, Hoboken, 2013). ISBN 9781118060407
109. G. Armelles, A. Cebollada, A. García-Martín, M.U. González, Magnetoplasmonics: combining magnetic and plasmonic functionalities. Adv. Opt. Mater. **1**(1), 10–35 (2013).
110. V.V. Temnov, Ultrafast acousto-magneto-plasmonics. Nat. Photonics **6**(11), 728–736 (2012).
111. P. Dombi, A. Hörl, P. Rácz, I. Márton, A. Trügler, J. R. Krenn, U. Hohenester, Ultrafast strong-field photoemission from plasmonic nanoparticles. Nano Lett. **13**(2), 674–678 (2013).
112. H. Gehan, C. Mangeney, J. Aubard, G. Lévi, A. Hohenau, J.R. Krenn, E. Lacaze, N. Félidj, Design and optical properties of active polymer-coated plasmonic nanostructures. J. Phys. Chem. Lett. **2**(8), 926–931 (2011).
113. R.A. Álvarez-Puebla, R. Contreras-Cáceres, I. Pastoriza-Santos, J. Pérez-Juste, L.M. Liz-Marzán, Au@pNIPAM colloids as molecular traps for surface-enhanced, spectroscopic, ultra-sensitive analysis. Angewandte Chemie International Edition **48**(1), 138–143 (2009).
114. Y. Fedutik, V.V. Temnov, O. Schops, U. Woggon, M.V. Artemyev, Exciton-plasmon-photon conversion in plasmonic nanostructures. Phys. Rev. Lett. **99**, 136802 (2007).
115. C. Kittel, *Introduction to Solid State Physics*, 8th edn. Wiley, New York, 2004) ISBN 978-0471415268
116. A. Arbouet, C. Voisin, D. Christofilos, P. Langot, N.D. Fatti, F. Vallée, J. Lermé, G. Celep, E. Cottancin, M. Gaudry, M. Pellarin, M. Broyer, M. Maillard, M.P. Pileni, M. Treguer, Electron-phonon scattering in metal clusters. Phys. Rev. Lett. **90**, 177401 (2003).

117. E.H. Hwang, R. Sensarma, S. Das Sarma, Plasmon-phonon coupling in graphene. Phys. Rev. B **82**(19), 195406 (2010).
118. R. Hillenbrand, T. Taubner, F. Keilmann, Phonon-enhanced light-matter interaction at the nanometre scale. Nature **418**(6894), 159–162 (2002).
119. U. Guler, V.M. Shalaev, A. Boltasseva, Nanoparticle plasmonics: going practical with transition metal nitrides. Mater. Today **18**(4), 227–237 (2015).
120. U. Guler, A.V. Kildishev, A. Boltasseva, V.M. Shalaev, Plasmonics on the slope of enlightenment: the role of transition metal nitrides. Faraday Discuss. **178**, 71–86 (2015).
121. A. Hohenau, H. Ditlbacher, B. Lamprecht, J.R. Krenn, A. Leitner, F.R. Aussenegg, Electron beam lithography, a helpful tool for nanooptics. Microelectron. Eng. **83**(4–9), 1464–1467 (2006).
122. M.R. Gonçalves, Plasmonic nanoparticles: fabrication, simulation and experiments. J. Phys. D: Appl. Phys. **47**(21), 213001 (2014).
123. B. Nikoobakht, M.A. El-Sayed, Preparation and growth mechanism of gold nanorods (NRs) using seed-mediated growth method. Chem. Mater. **15**, 1957–1962 (2003).
124. Q. Li, T. Bürgi, H. Chen, Preparation of gold nanorods of high quality and high aspect ratio. J. Wuhan Univer. Technology – Mater. Sci. Ed. **25**, 104–107 (2010).
125. J.-C. Tinguely, *The influence of nanometric surface morphology on surface plasmon resonances and surface enhanced effects of metal nanoparticles*. Ph.D. thesis, Institut für Physik, Karl-Franzens-Universität Graz, 2012.
126. L.M. Liz-Marzán, Nanometals: formation and color. Mater. Today **7**(2), 26–31 (2004).
127. A.R. Siekkinen, J.M. McLellan, J. Chen, Y. Xia, Rapid synthesis of small silver nanocubes by mediating polyol reduction with a trace amount of sodium sulfide or sodium hydrosulfide. Chem. Phys. Lett. **432**(4–6), 491–496 (2006).
128. Y. Sun, Y. Xia, Triangular nanoplates of silver: synthesis, characterization, and use as sacrificial templates for generating triangular nanorings of gold. Adv. Mat. **15**(9), 695–699 (2003).
129. D. Koller, *Luminescence coupling to dielectric and metallic nanostructures*. Ph.D. thesis, Institut für Physik, Karl-Franzens-Universität Graz, 2009.

Chapter 3
Theory

The work of James Clerk Maxwell changed the world forever

ALBERT EINSTEIN

The unification of the theories describing electric and magnetic aspects of our world was one of the great scientific achievements in the nineteenth century [1–3] and brought us a very successful part of theoretical physics: classical field theory. A detailed overview about the historical evolution from René Descartes up to Maxwell and Lorentz can be found in Whittaker [4]. After the revolution of our understanding of the basic forces and constituents of matter in the last 100 years, classical electrodynamics found its place in a sector of the unified description of particles and interactions known as the standard model [5].

3.1 Quantum Versus Classical Field Theory

Atoms and their corresponding electromagnetic fields fluctuate quite rapidly on the nanoscale, so usually we need to average over a larger region to obtain a macroscopic theory. In this sense the concept of the ordinary electromagnetic fields is a classical notion. It can be thought of as the classical limit (limit of large photon numbers and small momentum and energy transfers) of quantum electrodynamics (QED)[1] [5]. But nanoparticles are situated in the gray zone between the micro- and macrocosm–they are very small compared to classical objects but they still consist of several thousands to millions of atoms. Nevertheless surface plasmons are bosonic quasiparticles and have a true quantum nature that has been demonstrated by tunneling experiments for example, see [7]. Hence, for the theoretical description we can either come from the bottom and try to apply a quantum mechanical

---

[1] A good introduction into the topic of QED can be found in [6] for example.

© Springer International Publishing Switzerland 2016
A. Trügler, *Optical Properties of Metallic Nanoparticles*, Springer Series
in Materials Science 232, DOI 10.1007/978-3-319-25074-8_3

treatment (a recent review about *quantum plasmonics* can be found in [8]) or we can deal with plasmonic structures in terms of classical field theory (and hope that the particles are not too small).

At what point is it justified that we neglect (or at least gloss over) the discrete photon aspect of the electromagnetic field and change from QED to Maxwell's theory?[2] In the domain of macroscopic phenomena the answer is virtually always, as the examples discussed in [5] elucidate: The root mean square electric field one meter away from a 100 W light bulb is of the order of 50 V/m and there are of the order of 10^{15} visible photons per cm^2 per second. Similarly, an antenna that emits isotropically with a power of 100 W at 10^8 Hz produces a root mean square electric field of only 0.5 mV/m at a distance of 100 km, but this still corresponds to a flux of 10^{12} photons per cm^2 per second. Ordinarily an apparatus will not be sensitive to the individual photons; the cumulative effect of many photons emitted or absorbed will appear as a continuous, macroscopically observable response. Then a completely classical description in terms of the Maxwell equations is permitted and is appropriate. In this sense, a rough estimate for the justification of a classical treatment is given by a high number of involved photons where at the same time their momentum has to be small compared to the material system.[3] This is true for metallic nanoparticles and in linear response, one can employ the *fluctuation-dissipation theorem* to relate the dielectric response to the dyadic Green tensor of Maxwell's theory where all the details of the metal dynamics are embodied in the dielectric function. This is exactly what we are going to do, we will hide the quantum-mechanical properties of matter in their dielectric description which is obtained by experiment (see Fig. 2.3 and [13]). In this way, we are communicating with the microscopic world via ε and μ.

Nevertheless, the entire concept of a dielectric function becomes questionable if the investigated nanoparticles are too small (caution for structures below 5 nm may be justified, see Chap. 8). Also if coupled particles get very close to each other, the onset of screening effects and electron tunneling across the gap region significantly modifies the optical response as reported in [14]. In the cited work, the authors present a fully quantum mechanical description of nanoparticle dimers in terms of time-dependent density functional theory and state that quantum effects for dimers become important for dimer separations below 1 nm.

---

[2]*Decoherence* is the keyword when it comes to the transition from the quantum to the classical world, an excellent review about that concept can be found in [9], for example (also see [10]). A very interesting debate about the meaning of quantum mechanics has been published by the same author in [11], by the way, where seventeen physicists and philosophers, all deeply concerned with understanding quantum mechanics, share their opinion on what comes next and how to make sense of the theory's strangeness.

[3]Because of energy and momentum conservation at least the time averaged electromagnetic field can still be treated in a classical way, the same goes for clearly quantum mechanical processes like spontaneous emission [5]. A detailed discussion of this topic is also given in the introductory chapters of [12].

Additional remark

If we neglect the quantum mechanical nature of atoms and think of them as spheres in a crystal lattice (an engrained and handy picture after all), we can make a rough and simple estimate about the number of atoms that are contained in an ordinary metallic nanoparticle. Most metals that are typically used for nanostructures have a face centered cubic (fcc) crystal structure, e.g. gold, silver or aluminum (Fig. 3.1):

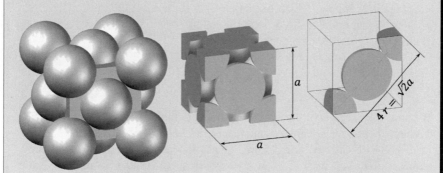

Fig. 3.1 Schematic unit cell of a fcc crystal with lattice constant a. The single cell without any atoms in neighboring volumes is shown in the middle

In this structure the atoms are most closely packed along the face diagonal, so we deduce the relation

$$4r = \sqrt{2}a \quad \Longrightarrow \quad a = \frac{4}{\sqrt{2}}r = 2\sqrt{2}r.$$

If we count the number of atoms in the unit cell, we obtain for the eight edge and six face atoms $8 \times 1/8 + 6 \times 1/2 = 4$ atoms per unit cell. The volume of one cell is given by $V_{uc} = a^3 = 16\sqrt{2}r^3$ and the volume of the contained atoms by $V_{atom} = 4 \times \frac{4\pi}{3}r^3$. The final packing density for fcc metals is then given by

$$\frac{V_{atom}}{V_{uc}} = \frac{\pi}{3\sqrt{2}} \approx 0.74 = 74\,\%. \tag{3.1}$$

This number also entered the annals of mathematics in form of the *Kepler conjecture* [15, 16], which claims that fcc packing is the best we can do if we

Additional remark

want to pack equally sized spheres[4] with the highest average density in three-dimensional Euclidean space.

To finally calculate the number of atoms in a metallic nanocluster we additionally need to know the size of one atom of the considered material. But that is a tricky task of course since atoms do not have sharp boundaries–still, the best way is again to assume an imaginary hard sphere with the so-called *van der Waals radius*, which gives 0.166 nm for gold, 0.172 nm for silver or 0.184 nm for aluminum. So we end up with the final estimate for the number of atoms in a metallic nanoparticle with volume V:

$$\frac{V}{V_{\text{atom}}} \cdot 0.74 \approx \frac{V}{r^3} \cdot 0.177. \tag{3.2}$$

We have also neglected the symmetry break at the surface of a nanoparticle, but nevertheless we are able to obtain a rough feeling for the number of atoms involved: A gold sphere with 25 nm in diameter consists of more than 300,000 atoms, a silver slab with $15 \times 40 \times 10$ nm^3 of about 200,000 atoms.

---

[4]Without the mathematical disguise the spheres took the less charming form of canon balls on a war vessel in the original seventeenth-century formulation. C. F. Gauss introduced the first attempts of a formal proof of this conjecture in 1831, but it was not until 1988 that the mathematician T. C. Hales was able to finally proof the conjecture by computational methods [17] (it was published after a 7 year long review process).

3.2 Maxwell's Theory of Electromagnetism

We treat the electromagnetic fields as three-dimensional vector fields. Such fields are fully determined by their divergence and curl–that explains the structure and appearance of their mathematical description: Maxwell's equations. In SI units and in their macroscopic version they read as [5, 18]

Macroscopic Maxwell equations

$$\nabla \cdot D(r, t) = \rho(r, t), \qquad \text{(Gauss's Law)} \qquad (3.3a)$$

$$\nabla \cdot B(r, t) = 0, \qquad \text{(magnetic analogon)} \qquad (3.3b)$$

$$\nabla \times H(r, t) = j(r, t) + \frac{\partial D(r, t)}{\partial t}, \qquad \text{(Ampère's Circuital Law)} \qquad (3.3c)$$

$$\nabla \times E(r, t) = -\frac{\partial B(r, t)}{\partial t}. \qquad \text{(Faraday's Induction Law)} \qquad (3.3d)$$

Here $B = \mu H$ is the magnetic field (magnetic permeability $\mu \approx \mu_0$ at optical frequencies, see Scct. 2.8), $D = \varepsilon E$ is the dielectric displacement,[5] ϱ the free charge density, and j the current density. In vacuum we would obtain $B = \mu_0 H$ and $D = \varepsilon_0 E$ with the vacuum permeability μ_0 and vacuum permittivity ε_0.

Maxwell's equations are partial differential equations of the first order. In many cases they are linear[6] in the fields E and B. Because of this linearity it is sufficient to only investigate time harmonic fields, any complex solution of the system can then be described as a superposition of them. Henceforth we will use

$$E(r, t) = E(r)e^{-i\omega t}, \qquad B(r, t) = B(r)e^{-i\omega t}. \qquad (3.4)$$

Additional remark

The inverse square law of the electrostatic force was shown quantitatively in experiments by Coulomb and Cavendish [5]. Applying the divergence theorem together with Gauss's law allows the derivation of the first of Maxwell's equations, Eq. (3.3a). But Coulomb's inverse square law also leads to another

---

[5]This relation connects the microscopic response with a macroscopic field, a more detailed discussion follows in Sect. 3.2.3.

[6]Nonlinear effects may arise at interaction with fiber glass or certain magnetic materials and many other systems.

Additional remark

remarkable condition: The photon has to be a massless particle [5]. We can verify this hypothesis solely with experiments, and Maxwell's equations are based on this assumption. The consequences of a massive photon are once again discussed in [5], for example, and the experimental verification of Coulomb's law already gives a very good upper limit for the photon mass m_γ, see [19]. Very accurate results for m_γ can be obtained by measuring the magnetic field of earth [20], viz.

$$m_\gamma < 4 \times 10^{-51} \text{ kg,} \qquad (3.5)$$

or the cosmic magnetic vector potential, see [21]. Also the *Schumann resonances* [22–25] (stationary electromagnetic waves along the circumference of the earth) allow for a very simple but surprisingly accurate estimation of the upper limit of m_γ. Earth and the ionosphere form a lossy cavity resonator and lightning discharges are constantly exciting Schumann resonances therein (thunderstorms around the globe produce approximately 50 lightning events per second [26]). The lowest Schumann resonance frequency is $\approx 8\,\text{Hz}$ and with Einstein's relation $\hbar\omega = 2\pi\hbar\nu = m_\gamma c^2$ we obtain $m_\gamma < 6 \times 10^{-50}\,\text{kg.}$[7]

From Gauss's law in Maxwell's equations it follows immediately that the magnetic field is purely transverse $(B_\parallel = 0)$ and the longitudinal part of E is related to the free charge distribution. By definition, a longitudinal or transverse vector field is characterized by the following relations (first expression in real space, second in reciprocal space; for simplicity we use the same symbols in both representations):

longitudinal field: $\qquad \nabla \times V_\parallel = 0, \qquad i k \times V_\parallel = 0,$

transverse field: $\qquad \nabla \cdot V_\perp = 0, \qquad i k \cdot V_\perp = 0.$

In this sense we obtain a clear geometrical meaning of longitudinal or transverse fields in the reciprocal space, they are either parallel or perpendicular to all k [6]. If no sources are present $(\rho = 0, j = 0)$ Maxwell's equations reduce to

---

Maxwell's equations in vacuum

$$\nabla \cdot D = 0, \qquad\qquad \nabla \times H - \frac{\partial D}{\partial t} = 0, \qquad (3.6a)$$

---

[7] $1\,\text{eV} = 1.602176565 \times 10^{-19}\,\text{kg}\,\text{m}^2/\text{s}^2$, see [27].

$$\nabla \cdot \boldsymbol{B} = 0, \qquad\qquad \nabla \times \boldsymbol{E} + \frac{\partial \boldsymbol{B}}{\partial t} = 0. \qquad\qquad (3.6b)$$

Additional remark

Maxwell's equations form the backbone of modern communication technology and have changed our world forever. They represent the rigorous mathematical abstraction of experimental findings, but they cannot be derived from any deeper principle of our universe till this day. Although some elegant attempts exist (e.g. based on the gauge invariance of classical mechanics [28] or quantum mechanics [29], or an extended Helmholtz theorem [30]), one always has to postulate certain relations or requirements to allow the derivation.[8]

Despite the fact that the wonderful symmetry of Eq. (3.6) permits a glimpse of the formal and simple beauty of nature and of the stunning mutual entanglement of the electromagnetic fields, Maxwell's equations also presciently anticipate the constancy of the speed of light as well as the principle of relativity (for an excellent guided tour through the historical development and radical concepts of the theory of relativity consult the still superior book of Max Born [32], for example). Hence, especially based on the work of Lorentz and Poincaré, special relativity emerged from electromagnetism and Einstein merely released the ideas from this derivation and put them on fundamental and philosophical ground.

3.2.1 Boundary Conditions at Interfaces of Different Media

All the interesting phenomena in plasmonics happen at the boundary of different media, thus the corresponding constraints and boundary conditions are essential. Fortunately we do not have to introduce additional equations at the boundary, everything is already contained in Maxwell's theory. All we have to do is to use

---

[8]Also see comments to Feynman's unpublished attempt in [31], where he mixes classical and quantum mechanical concepts by starting with Newton's law of motion and implying the commutation relation between position and velocity. The discussion of a heuristic derivation based on symmetry, charge conservation, superposition, the existence of electromagnetic waves and the Lorentz force can be found in [18], for example.

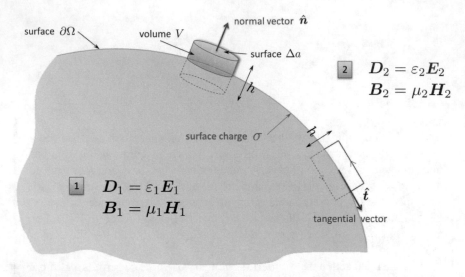

Fig. 3.2 Schematic diagram of the interface between two different dielectric media 1 and 2, based on [5]. The small cylinder of height h is given by the volume V. The cylinder's lower and upper top surface area is labeled by Δa. The normal vector $\hat{n}$ points in outside direction (from medium 1 to medium 2). The vector $\hat{t}$ is tangential to the surface $\partial\Omega$. The surface spanned by the rectangular kink is perpendicular to the interface, i.e., $\hat{t}$ is perpendicular to the kink. At the interface $\partial\Omega$ exists an idealized surface charge density $\bar{\sigma}$

Gauss' law (3.3a) and to apply Stokes' theorem[9] on Faraday's law (3.3d). By following [5], let V be a finite space volume bordered by the surface area $\partial\Omega$, and let $\hat{n}$ be the unit vector in outside direction of the surface element da (see Fig. 3.2).

At first we transform Eqs. (3.3a) and (3.3b) to their corresponding integral form and apply them to the volume of the small cylinder in Fig. 3.2. In the limit of infinitesimal height ($h \to 0$) only the lower and upper top surface area are non-zero. If we assume that this surface has the value Δa and approximately set D_1 and D_2 constant within the surface element, it follows that

$$\oint_S \boldsymbol{D} \cdot \hat{n} \, da = (\boldsymbol{D}_2 - \boldsymbol{D}_1) \cdot \hat{n} \, \Delta a, \tag{3.7}$$

$$\oint_S \boldsymbol{B} \cdot \hat{n} \, da = (\boldsymbol{B}_2 - \boldsymbol{B}_1) \cdot \hat{n} \, \Delta a. \tag{3.8}$$

---

[9]Stokes' theorem

$$\int_\Omega (\boldsymbol{\nabla} \times \boldsymbol{F}) \cdot \hat{n} \, dS = \oint_{\partial\Omega} \boldsymbol{F} dr,$$

relates the curl of a vector field $\boldsymbol{F}$ integrated over a surface to the line integral of the vector field at the boundary. Written in a more general formulation it also contains the divergence theorem or Green's theorem as special cases.

If the charge density ϱ at the surface $\partial\Omega$ is singular and forms an idealized surface charge density $\bar\sigma$,[10] the right hand side of Eq. (3.3a) yields

$$\int_V \varrho\,\mathrm{d}V = \int_V \bar\sigma\,\delta(\boldsymbol{r}-\boldsymbol{s})\,\mathrm{d}V = \int_{\partial V} \bar\sigma\,\mathrm{d}a = \bar\sigma\,\Delta a. \qquad (3.9)$$

Thus the boundary conditions for the normal components of $\boldsymbol{D}$ and $\boldsymbol{B}$ follow as:

$$(\boldsymbol{D}_2 - \boldsymbol{D}_1)\cdot\hat{\boldsymbol{n}} = \bar\sigma, \qquad (3.10a)$$

$$(\boldsymbol{B}_2 - \boldsymbol{B}_1)\cdot\hat{\boldsymbol{n}} = 0. \qquad (3.10b)$$

Applying a similar procedure to the rectangular kink C with Eqs. (3.3c) and (3.3d), using Stokes' theorem yields the boundary conditions for the tangential field components:

$$\hat{\boldsymbol{n}} \times (\boldsymbol{E}_2 - \boldsymbol{E}_1) = 0, \qquad (3.11a)$$

$$\hat{\boldsymbol{n}} \times (\boldsymbol{H}_2 - \boldsymbol{H}_1) = \bar{\boldsymbol{h}}, \qquad (3.11b)$$

where $\bar{\boldsymbol{h}}$ is an idealized surface current density. For the quasistatic case the scalar potential ϕ obeys analogous conditions, which are quickly derived by using the relations

$$\boldsymbol{D}_1 = \varepsilon_1\boldsymbol{E}_1 = -\varepsilon_1\nabla\phi_1, \qquad \boldsymbol{D}_2 = \varepsilon_2\boldsymbol{E}_2 = -\varepsilon_2\nabla\phi_2. \qquad (3.12)$$

Inserting this in Eq. (3.10) leads to

$$\varepsilon_1\phi_1'\Big|_{\text{surf}} = \varepsilon_2\phi_2'\Big|_{\text{surf}}, \qquad (3.13)$$

where the apostrophe $'$ denotes the normal derivative, i.e. the derivative in a direction perpendicular to the surface. Equation (3.11) shows that the tangential components of $\boldsymbol{E}$ are continuous at the interface and therefore the derivation of the potential in tangential direction is also continuous.

---

[10]Since a redefinition of the symbol σ and j will become necessary later on, the additional bar over the symbols has been introduced to avoid confusions.

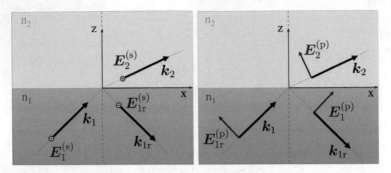

Fig. 3.3 Reflection and refraction at a planar interface. The *left panel* shows a TE- or (s)-polarized plane wave, the *right panel* a TM- or (p)-polarized wave

Using spherical coordinates and considering a sphere with radius a and surface charge density $\bar{\sigma}$ in a dielectric medium yields

$$\varepsilon_1 \frac{\partial \phi_1}{\partial r}\bigg|_a - \varepsilon_2 \frac{\partial \phi_2}{\partial r}\bigg|_a = \bar{\sigma}, \tag{3.14a}$$

$$\frac{\partial \phi_1}{\partial \theta}\bigg|_a - \frac{\partial \phi_2}{\partial \theta}\bigg|_a = 0. \tag{3.14b}$$

3.2.2 Fresnel Coefficients

The most simple scenario where boundary conditions come into play is given by an interface that divides space into two separate regions. Let us assume that a plane wave is impinging at such a boundary and that medium 1 and medium 2 are two different dielectric materials with refractive indices n_1 and n_2, see Fig. 3.3. The French physicist Augustin-Jean Fresnel[11] solved the problem in the nineteenth century and deduced reflection and transmission coefficients for the electromagnetic wave. Here we just briefly depict the summary presented in [33], a more detailed derivation of Fresnel's equations can be found in [34] for example. The expression $E_1 e^{i k_1 \cdot r - i\omega t}$ describes an arbitrarily polarized plane wave propagating in k_1-direction in medium 1. It can always be written as a superposition of two plane waves polarized parallel and perpendicular to the plane of incidence

$$E_1 = E_1^{(s)} + E_1^{(p)}. \tag{3.15}$$

---

[11]Born 10th May 1788 in Broglie, Eure (Haute-Normandie); † 14th July 1827 in Ville-d'Avray, Hauts-de-Seine.

Now $E_1^{(s)}$ is parallel to the interface separating medium 1 and 2 while $E_1^{(p)}$ is perpendicular to the wavevector k and $E_1^{(s)}$, see Fig. 3.3. The transmission or reflection of E_1 at the interface does not change the polarization, i.e. (s) and (p) are conserved. The transverse components of the wavevector are also conserved at the boundary and we get

$$k_1 = (k_x, k_y, k_{z_1}), \quad |k_1| = k_1 = n_1 \frac{\omega}{c} = \sqrt{\varepsilon_1 \mu_1} \omega, \tag{3.16}$$

$$k_2 = (k_x, k_y, k_{z_2}), \quad |k_2| = k_2 = n_2 \frac{\omega}{c} = \sqrt{\varepsilon_2 \mu_2} \omega. \tag{3.17}$$

The longitudinal components are given by

$$k_{z_1} = \sqrt{k_1^2 - (k_x^2 + k_y^2)}, \quad k_{z_2} = \sqrt{k_2^2 - (k_x^2 + k_y^2)}. \tag{3.18}$$

The expressions for the field amplitudes of the reflected and transmitted waves finally follow from the boundary conditions and are given by

$$E_{1r}^{(s)} = r^{(s)}(k_x, k_y)E_1^{(s)}, \qquad E_{1r}^{(p)} = r^{(p)}(k_x, k_y)E_1^{(p)},$$

$$E_2^{(s)} = t^{(s)}(k_x, k_y)F_1^{(s)}, \qquad E_2^{(p)} = t^{(p)}(k_x, k_y)E_1^{(p)},$$

where the Fresnel coefficients r and t are defined as [33]

Fresnel coefficients

$$r^{(s)}(k_x, k_y) = \frac{\mu_2 k_{z_1} - \mu_1 k_{z_2}}{\mu_2 k_{z_1} + \mu_1 k_{z_2}}, \qquad r^{(p)}(k_x, k_y) = \frac{\varepsilon_2 k_{z_1} - \varepsilon_1 k_{z_2}}{\varepsilon_2 k_{z_1} + \varepsilon_1 k_{z_2}}, \tag{3.19a}$$

$$t^{(s)}(k_x, k_y) = \frac{2\mu_2 k_{z_1}}{\mu_2 k_{z_1} + \mu_1 k_{z_2}}, \qquad t^{(p)}(k_x, k_y) = \frac{2\varepsilon_2 k_{z_1}}{\varepsilon_2 k_{z_1} + \varepsilon_1 k_{z_2}} \sqrt{\frac{\mu_2 \varepsilon_1}{\mu_1 \varepsilon_2}}. \tag{3.19b}$$

The coefficients depend on the longitudinal wavenumbers k_{z_1} and k_{z_2} which can be expressed in terms of the angle of incidence θ.

3.2.3   Linear and Nonlinear Optical Response

If we apply an external electric field to a polarizable (dielectric) medium, the electrons in the material response with a microscopic shift but still remain bound to their associated atoms. The cumulative effect of all displaced electrons results in a macroscopic polarization of the material that can be described by a net charge distribution. In terms of Maxwell's theory it becomes useful if we distinguish between this bound charge distribution and that of free charges, which we have introduced as ϱ in Eq. (3.3a). The contribution of the bound charges is incorporated elsewhere, but let us discuss this in more detail. The electromagnetic fields obey the following relations

$$D = \varepsilon_0 E + P, \qquad H = \frac{1}{\mu_0} B - M, \qquad (3.20)$$

where P is the dipole moment per unit volume and M refers to the magnetic moment per unit volume. Since we focus our attention only to nonmagnetic media, we can set $M \equiv 0$ (see Sect. 2.8) and combine Maxwell's equations to

$$\nabla^2 E - \nabla(\nabla \cdot E) - \frac{1}{c^2} \frac{\partial^2 E}{\partial t^2} - \frac{1}{\varepsilon_0 c^2} \frac{\partial^2 P}{\partial t^2} = 0, \qquad (3.21)$$

by taking the curl on Eq. (3.3d) and inserting Eq. (3.3c). The speed of light in vacuum is given by $c = \sqrt{1/\mu_0 \varepsilon_0}$. In principle, one now requires a full microscopic theory of the response of a particular material to relate the macroscopic electric field E to the polarization P [5, 35]. Making some assumptions about the relationship between P and E will make our lives much easier.

The change in electrostatic potential over distances of the order of an angstrom can be several electron volts. In this sense, an electron bound to an atom or molecule, or moving through a solid or dense liquid, experiences electric fields of the order of 10^9 V/cm [35]. The laboratory fields of interest are then small compared to the electric fields experienced by the electrons in the atoms and molecules of the matter under investigation. In this circumstance, we can expand $P(r, t)$ in a Taylor series in powers of the macroscopic field[12] $E(r, t)$. The αth Cartesian component of the dipole moment per unit volume is a function of the three Cartesian components of the electric field $E_\beta = E_\beta(r, t)$, with $\beta \in \{x, y, z\}$. Therefore we can write the Taylor series as

$$\frac{P_\alpha(r, t)}{\varepsilon_0} = P_\alpha^{(0)} + \sum_\beta \left(\frac{\partial P_\alpha}{\partial E_\beta} \right)_0 E_\beta + \frac{1}{2!} \sum_{\beta\gamma} \left(\frac{\partial^2 P_\alpha}{\partial E_\beta \partial E_\gamma} \right)_0 E_\beta E_\gamma +$$

$$+ \frac{1}{3!} \sum_{\beta\gamma\delta} \left(\frac{\partial^3 P_\alpha}{\partial E_\beta \partial E_\gamma \partial E_\delta} \right)_0 E_\beta E_\gamma E_\delta + \cdots .$$

Here we have assumed that the dipole moment $P(r, t)$ depends on the electric field E at the same point r in space and the same time t, which is not really a realistic

---

[12]Another method sometimes discussed in literature models the atomic or molecular structure explicitly. There one relates the dipole moment per unit volume to that of an atomic or molecular constituent and writes this as a Taylor series similar to our approach, see [35, 36] and references therein.

assumption–we will introduce a more proper treatment in the next section and we will see that we can incorporate certain nonlocal aspects in the susceptibility tensor.

In plasmonics we are usually interested in dielectric materials within which any dipole moment is induced by the external excitation. Therefore the electric dipole moment per unit volume at zero field, $P_\alpha^{(0)}$, vanishes,[13] and we will henceforth write

$$\frac{P_\alpha(\boldsymbol{r},t)}{\varepsilon_0} = \sum_\beta \chi_{\alpha\beta}^{(1)} E_\beta + \sum_{\beta\gamma} \chi_{\alpha\beta\gamma}^{(2)} E_\beta E_\gamma + \sum_{\beta\gamma\delta} \chi_{\alpha\beta\gamma\delta}^{(3)} E_\beta E_\gamma E_\delta + \cdots, \quad (3.22)$$

where the susceptibilities $\chi^{(i)}$ are tensors of $(i+1)$th rank. $\chi^{(1)}$ is the ordinary susceptibility of dielectric theory (usually a diagonal matrix) and $\chi^{(2)}$, $\chi^{(3)}$ are referred to as the second and third order susceptibilities, respectively. Now we can decompose the dipole moment into a part which is linear in the electric field, and one part which is nonlinear:

$$P_\alpha(\boldsymbol{r},t) = P_\alpha^{(L)}(\boldsymbol{r},t) + P_\alpha^{(NL)}(\boldsymbol{r},t), \quad (3.23)$$

where

Linear and nonlinear dipole moment per unit volume

$$P_\alpha^{(L)}(\boldsymbol{r},t) = \sum_\beta \varepsilon_0 \chi_{\alpha\beta}^{(1)} E_\beta, \quad (3.24a)$$

$$P_\alpha^{(NL)}(\boldsymbol{r},t) = \sum_{\beta\gamma} \varepsilon_0 \chi_{\alpha\beta\gamma}^{(2)} E_\beta E_\gamma + \sum_{\beta\gamma\delta} \varepsilon_0 \chi_{\alpha\beta\gamma\delta}^{(3)} E_\beta E_\gamma E_\delta + \cdots. \quad (3.24b)$$

With the electric susceptibility χ_e (simplified symbol instead of $\chi^{(1)}$) we now obtain the previously discussed relation, where the microscopic response is incorporated in the dielectric function:

$$\boldsymbol{D} = \varepsilon_0 \boldsymbol{E} + \boldsymbol{P} = \varepsilon_0(1 + \chi_e)\boldsymbol{E} = \varepsilon \boldsymbol{E}. \quad (3.25)$$

Furthermore we now have a clear distinction between linear and nonlinear optics: If we insert $P_\alpha^{(L)}(\boldsymbol{r},t)$ into Maxwell's equations we obtain a description of electromagnetic wave propagation in (possibly crystalline) media, described by an electric

---

[13]The electrical analogs of ferromagnets, which possess a spontaneous magnetization per unit volume, are the so-called ferroelectrics. In these materials the dipole moment $\boldsymbol{P}^{(0)}$ in the absence of an electric field is nonzero and leads to the presence of a static, macroscopic electric field, $\boldsymbol{E}^{(0)}(\boldsymbol{r})$. Such time independent effects may be analyzed by the methods of electrostatics and can be accounted for by including an effective charge density $\varrho_p = -\boldsymbol{\nabla} \cdot \boldsymbol{P}^{(0)}$, for example.

susceptibility tensor $\chi_{\alpha\beta}$, in linear response. All nonlinear effects are part of the higher order susceptibilities.

Since $\boldsymbol{P}$ and $\boldsymbol{E}$ are vectors, and thus are odd under inversion symmetry, $\chi^{(2)}$ must vanish in any material that is left invariant in form under inversion.[14] But if the symmetry is broken (at the interface from one medium to another, for example) or in the case of surface imperfections [37], we also obtain $\chi^{(2)}$ contributions for centrosymmetric materials like gold or silver, see Sect. 7 and [38].

3.2.4 Nonlocal in Space and Time

The macroscopic field $\boldsymbol{E}(\boldsymbol{r}, t)$ acts as a driving field that leads to a rearrangement of the electrons and nuclei in the material. The result is the induced dipole moment $\boldsymbol{P}$, which of course will not be built up instantaneously, but is instead the consequence of the response of the system over some characteristic time interval $t - t' > 0$ in the recent past. If, on the other hand, we consider an incident electric field well localized in space, it will lead to an electronic rearrangement in a certain small region of the material. Because of the interaction with neighboring constituents, the material gets polarized in the vicinity of the excitation as well. It follows then that the dipole moment $\boldsymbol{P}(\boldsymbol{r}, t)$ depends not only on the field at time t and position $\boldsymbol{r}$, but must be written as a convolution in space and time (exemplified only for linear response)

$$P_\alpha^{(L)}(\boldsymbol{r}, t) = \sum_\beta \int_{\mathbb{R}^3 \otimes \mathbb{R}} \mathrm{d}^3 r' \, \mathrm{d}t' \, \varepsilon_0 \chi_{\alpha\beta}^{(1)}(\boldsymbol{r} - \boldsymbol{r}', t - t') E_\beta(\boldsymbol{r}', t'). \tag{3.26}$$

If there are no variations in density or composition, the medium can be treated as homogeneous in nature. Then the susceptibility tensor χ will not depend on $\boldsymbol{r}$ or $\boldsymbol{r}'$ separately but only on the spatial difference $\boldsymbol{r} - \boldsymbol{r}'$. A second simplification can be exploited. If the electric field exhibits only a slow variation in space and time we may use $\boldsymbol{E}(\boldsymbol{r}', t') \approx \boldsymbol{E}(\boldsymbol{r}, t)$ and recover Eq. (3.24a), where we now have shown the structure of the susceptibility tensor in more detail:

$$\chi_{\alpha\beta}^{(1)}(\boldsymbol{r}, t) = \int \mathrm{d}^3 r' \, \mathrm{d}t' \, \chi_{\alpha\beta}(\boldsymbol{r} - \boldsymbol{r}', t - t'). \tag{3.27}$$

The physical meaning of nonlinear response of a material becomes clear if analyzed in Fourier space. Therefore we will briefly list the basic Fourier decompositions, again exemplified for linear response. For simplicity we will use the same symbols for functions in Fourier as well as in real space and follow the notation in [35]. The

---

[14]This is the case for metals like Au or Ag, for the semiconductors Si and Ge as well as for liquids, gases, and for a number of other common crystals. The interested reader may find a very useful compilation of the nonzero elements of $\chi^{(2)}$ and $\chi^{(3)}$ for crystals of various symmetry in [36].

Fourier transform operator and its inverse are denoted by the symbols $\mathcal{F}$ and $\mathcal{F}^{-1}$, respectively.

$$E_\beta(\mathbf{k}, \omega) = \mathcal{F}[E_\beta(\mathbf{k}, \omega)] \equiv \int \mathrm{d}^3 r \, \mathrm{d}t \, E_\beta(\mathbf{r}, t) \mathrm{e}^{-\mathrm{i}\mathbf{k}\cdot\mathbf{r}} \mathrm{e}^{\mathrm{i}\omega t} \qquad (3.28\mathrm{a})$$

$$E_\beta(\mathbf{r}, t) = \mathcal{F}^{-1}[E_\beta(\mathbf{k}, \omega)] \equiv \int \frac{\mathrm{d}^3 k \, \mathrm{d}\omega}{(2\pi)^4} E_\beta(\mathbf{k}, \omega) \mathrm{e}^{\mathrm{i}\mathbf{k}\cdot\mathbf{r}} \mathrm{e}^{-\mathrm{i}\omega t}. \qquad (3.28\mathrm{b})$$

With these definitions the transformation for $P_\alpha^{(L)}$ follows directly from Eq. (3.26):

$$P_\alpha^{(L)}(\mathbf{r}, t) = \int \frac{\mathrm{d}^3 k \, \mathrm{d}\omega}{(2\pi)^4} P_\alpha^{(L)}(\mathbf{k}, \omega) \mathrm{e}^{\mathrm{i}\mathbf{k}\cdot\mathbf{r}} \mathrm{e}^{-\mathrm{i}\omega t}, \qquad (3.29)$$

where

$$P_\alpha^{(L)}(\mathbf{k}, \omega) = \sum_\beta \varepsilon_0 \chi_{\alpha\beta}^{(1)}(\mathbf{k}, \omega) E_\beta(\mathbf{k}, \omega),$$

$$\chi_{\alpha\beta}^{(1)}(\mathbf{k}, \omega) = \int \mathrm{d}^3 r \, \mathrm{d}t \, \chi_{\alpha\beta}^{(1)}(\mathbf{r}, t) \mathrm{e}^{-\mathrm{i}\mathbf{k}\cdot\mathbf{r}} \mathrm{e}^{\mathrm{i}\omega t}.$$

For the rest of this chapter we will stick to the linear response and assume the simple linear proportionality between $\mathbf{P}$ and $\mathbf{E}$. Nonlinear optical responses will again be discussed in Chap. 7.

3.2.5 Electromagnetic Potentials

We now return to Maxwell's equations. Let us recall their appearance in a source free frequency space:

$$\boldsymbol{\nabla} \cdot \mathbf{D}(\mathbf{r}, \omega) = 0, \qquad \boldsymbol{\nabla} \times \mathbf{H}(\mathbf{r}, \omega) + \mathrm{i}\omega \mathbf{D}(\mathbf{r}, \omega) = \mathbf{0}, \qquad (3.30\mathrm{a})$$

$$\boldsymbol{\nabla} \cdot \mathbf{B}(\mathbf{r}, \omega) = 0, \qquad \boldsymbol{\nabla} \times \mathbf{E}(\mathbf{r}, \omega) - \mathrm{i}\omega \mathbf{B}(\mathbf{r}, \omega) = \mathbf{0}. \qquad (3.30\mathrm{b})$$

Taking the curl on Ampère's and Faraday's law respectively and substituting the corresponding equations leads us to the wave equation of Helmholtz form:

Wave equation for electromagnetic fields

$$\left(\nabla^2 + \mu\varepsilon\omega^2\right) \begin{Bmatrix} \mathbf{E}(\mathbf{r}, \omega) \\ \mathbf{B}(\mathbf{r}, \omega) \end{Bmatrix} = \left(\nabla^2 + n^2 \frac{\omega^2}{c^2}\right) \begin{Bmatrix} \mathbf{E}(\mathbf{r}, \omega) \\ \mathbf{B}(\mathbf{r}, \omega) \end{Bmatrix} = \mathbf{0}, \qquad (3.31)$$

where we have used the refractive index $n = \sqrt{\mu\varepsilon/\mu_0\varepsilon_0}$ to uncover the emergence of the speed of light $c = \sqrt{1/\mu_0\varepsilon_0}$. This equation is a central result of Maxwell's theory, since it postulates the existence of electromagnetic waves.[15] Thus as a possible solution we can write down a plane wave propagating in $\hat{e}_r$-direction: $e^{i\boldsymbol{k}\cdot\boldsymbol{r}-i\omega t}$, with $\boldsymbol{k} = k\,\hat{e}_r$ and $k = n\frac{\omega}{c}$.

Additional remark

The appearance of the speed of light in the electromagnetic theory had an immense impact and is a noteworthy story. In 1676 the Danish astronomer Ole Rømer was the first to prove that the speed of light is finite. By observing the duration of the eclipses of the moons of Jupiter, he was able to give a surprisingly accurate first estimate of c. In 1820 Hans Cristian Ørsted, another Danish scientist, experimentally showed that magnetic needles are influenced by electric currents and in the same year Jean-Baptiste Biot and Félix Savart discovered the underlying quantitative law. In their original formulation occurred a proportionality constant, which was measured precisely by Weber and Kohlrausch in 1856 [32] and yielded $300,000\,km/s$. For the first time the deep connection between optics and electromagnetism emerged and finally Maxwell built the bridge between these two scientific domains [32].

If we change from first to second order equations, we can combine the four coupled expressions of Eq. (3.30) into two new equations and introduce the vector potential $\boldsymbol{A}$ and the scalar potential ϕ. The fields can then be expressed as

---

Electromagnetic fields expressed with potentials

$$\boldsymbol{B} = \nabla \times \boldsymbol{A}, \tag{3.32a}$$

$$\boldsymbol{E} = -\nabla\phi - \frac{\partial \boldsymbol{A}}{\partial t} = -\nabla\phi + i\omega\boldsymbol{A}. \tag{3.32b}$$

The differential equations for $\boldsymbol{A}$ and ϕ still form a coupled system, which can be derived in a heartbeat by inserting Eq. (3.32) into Maxwell's equations (3.3):

$$\nabla^2\phi + \frac{\partial}{\partial t}(\nabla \cdot \boldsymbol{A}) = -\frac{\rho}{\varepsilon},$$

$$\nabla^2\boldsymbol{A} - \mu\varepsilon\frac{\partial^2\boldsymbol{A}}{\partial t^2} - \nabla\left(\nabla \cdot \boldsymbol{A} + \mu\varepsilon\frac{\partial\phi}{\partial t}\right) = -\mu\boldsymbol{j}.$$

---

[15] It was Helmholtz' student Heinrich Hertz who subsequently provided proof for their existence in the laboratory through his famous experiments with oscillating charges and currents.

Because of the gauge invariance of the potentials we can choose A and ϕ in such a way that they fulfill the *Lorenz condition* [39]

$$\nabla \cdot A + \mu\varepsilon\frac{\partial\phi}{\partial t} = \nabla \cdot A - i\frac{\omega\mu_r\varepsilon_r}{c^2}\phi = 0. \tag{3.33}$$

This condition does not entirely fix the gauge, but it is coordinate independent (and therefore naturally fits into special relativity) and leads to two decoupled wave equations for A and ϕ that are completely equivalent to Maxwell's equations (3.3):

Helmholtz equation for potentials

$$\nabla^2\phi + k^2\phi = -\frac{\rho}{\varepsilon}, \tag{3.34a}$$

$$\nabla^2 A + k^2 A = -\mu j. \tag{3.34b}$$

The wavenumber in the corresponding medium is again given by $k = n\frac{\omega}{c} = \sqrt{\mu\varepsilon}\omega$.

Additional remark

Maxwell's equations interweave space and time, electric and magnetic fields in such a wonderful way that Boltzmann was moved to express his deepest admiration [1]. Throughout time the equations changed their appearance[16] and hence are a vivid example of the mathematical beauty and the huge amount of physics that can be contained in one single line (constants set to 1, table adopted from [1]):

Homogeneous equations	*Inhomogeneous equations*

Original form:

$$\frac{\partial B_x}{\partial x} + \frac{\partial B_y}{\partial y} + \frac{\partial B_z}{\partial z} = 0, \qquad \frac{\partial E_x}{\partial x} + \frac{\partial E_y}{\partial y} + \frac{\partial E_z}{\partial z} = \rho,$$

$$\frac{\partial E_z}{\partial y} - \frac{\partial E_y}{\partial z} = -\frac{\partial B_x}{\partial t} \qquad \frac{\partial B_z}{\partial y} - \frac{\partial B_y}{\partial z} = j_x + \frac{\partial E_x}{\partial t}$$

---

[16]The most common form as a set of four equations expressed in the language of vector calculus was independently proposed by Heaviside and Hertz as a concise version of Maxwell's original set of equations [18].

Additional remark

$$\frac{\partial E_x}{\partial z} - \frac{\partial E_z}{\partial x} = -\frac{\partial B_y}{\partial t} \qquad \frac{\partial B_x}{\partial z} - \frac{\partial B_z}{\partial x} = j_y + \frac{\partial E_y}{\partial t}$$

$$\frac{\partial E_y}{\partial x} - \frac{\partial E_x}{\partial y} = -\frac{\partial B_z}{\partial t} \qquad \frac{\partial B_y}{\partial x} - \frac{\partial B_x}{\partial y} = j_z + \frac{\partial E_z}{\partial t}$$

End of nineteenth century:

$$\operatorname{div} \boldsymbol{B} = 0 \qquad\qquad \operatorname{div} \boldsymbol{E} = \rho$$

$$\operatorname{rot} \boldsymbol{E} = -\dot{\boldsymbol{B}} \qquad\qquad \operatorname{rot} \boldsymbol{B} = \boldsymbol{j} + \dot{\boldsymbol{E}}$$

Beginning of twentieth century:

$$* F^{\beta\alpha}{}_{,\alpha} = 0 \qquad\qquad F^{\beta\alpha}{}_{,\alpha} = j^\beta$$

Mid of twentieth century:

$$dF = 0 \qquad\qquad \delta F = J$$

3.3 Kramers-Kronig Relations

The microscopic Maxwell equations are local in time, i.e. the electromagnetic fields, currents and charges all depend on the same time t. To allow a macroscopic description of nanoscale physics we have integrated over the microscopic degrees of freedom and introduced the frequency dependent response function $\varepsilon(\omega)$. What we have tacitly ignored thereby is that this leads to temporally nonlocal equations [5]. In Sect. 3.2.4 we have already addressed nonlocalities in space and time by arguing that there is no instantaneous response of a system but some characteristic time interval $t - t' > 0$ is required to build up induced electromagnetic fields. We will now see that a temporally nonlocal response is also a direct consequence of the frequency dependence of the dielectric function and that the consideration of causality in the system (viz. no action before the cause) leads to a very general connection of the real and imaginary part of ε.

Let us start with Eq. (3.25), which connects the microscopic response with the macroscopic dielectric displacement:

$$\boldsymbol{D}(\boldsymbol{r}, \omega) = \varepsilon(\omega)\boldsymbol{E}(\boldsymbol{r}, \omega).$$

By following the notation in [5], we treat the spatial coordinate as a parameter and write down the Fourier integrals with symmetric prefactors as

$$\boldsymbol{D}(\boldsymbol{r}, t) = \frac{1}{\sqrt{2\pi}} \int_{-\infty}^{\infty} \boldsymbol{D}(\boldsymbol{r}, \omega)e^{-i\omega t}\, d\omega, \quad \boldsymbol{D}(\boldsymbol{r}, \omega) = \frac{1}{\sqrt{2\pi}} \int_{-\infty}^{\infty} \boldsymbol{D}(\boldsymbol{r}, t)e^{i\omega t'}\, dt'.$$

If we now use Eq. (3.25) and substitute the Fourier representation of $\boldsymbol{E}(\boldsymbol{r},\omega)$ we get

$$\boldsymbol{D}(\boldsymbol{r}, t) = \frac{1}{2\pi} \int_{-\infty}^{\infty} d\omega\, \varepsilon(\omega)e^{-i\omega t} \int_{-\infty}^{\infty} dt'\, e^{i\omega t'}\boldsymbol{E}(\boldsymbol{r}, t'). \tag{3.35}$$

With $\varepsilon(\omega) = \varepsilon_0[1 + \chi_e(\omega)]$ and the Fourier representation of the Dirac delta function

$$\delta(t - t') = \frac{1}{2\pi} \int_{-\infty}^{\infty} d\omega\, e^{-i\omega(t-t')}, \tag{3.36}$$

we can rewrite Eq. (3.35) as

$$\boldsymbol{D}(\boldsymbol{r}, t) = \varepsilon_0 \left[\int_{-\infty}^{\infty} dt'\, \delta(t - t')\boldsymbol{E}(\boldsymbol{r}, t') + \frac{1}{2\pi} \int_{-\infty}^{\infty} dt' \int_{-\infty}^{\infty} d\omega\, \chi_e(\omega)e^{-i\omega(t-t')}\boldsymbol{E}(\boldsymbol{r}, t') \right],$$

where we have assumed that the orders of integration can be interchanged. After the substitution $\tau = (t - t')$ we end up with a connection between D and E that is nonlocal in time[17]

Causality relation between D and E in the time domain

$$D(r, t) = \varepsilon_0 \left[E(r, t) + \int\limits_0^\infty d\tau\, \mathcal{G}(\tau) E(r, t - \tau) \right],$$

(3.37)

with $\mathcal{G}(\tau)$ being the Fourier transform of χ_e

$$\mathcal{G}(\tau) = \frac{1}{2\pi} \int\limits_0^\infty d\omega \left[\frac{\varepsilon(\omega)}{\varepsilon_0} - 1 \right] e^{-i\omega\tau}.$$

(3.38)

The lower limit of the integration has been changed from $-\infty$ to 0 since we have $\mathcal{G} = 0$ for $\tau < 0$ (see discussion in [5] for example). If the dielectric function is independent of the frequency ω, the integral in Eq. (3.38) is proportional to the Dirac delta function again and Eq. (3.37) recovers the original instantaneous expression $D = \varepsilon E$. If on the other hand ε varies with ω, the dielectric displacement D at time t depends on the electric field E prior to that time, which of course stands for causality in the system (the system cannot squeal before it is hurt). In a nutshell we can summarize that even though Maxwell's equations are local in time, integrating over the microscopic degrees of freedom leads to nonlocal material equations (in Chap. 8 we will also discuss spatial nonlocality).

We can also use the susceptibility kernel (3.38) to express the dielectric function as

$$\frac{\varepsilon(\omega)}{\varepsilon_0} = 1 + \int\limits_0^\infty d\tau\, \mathcal{G}(\tau) e^{i\omega\tau},$$

(3.39)

which allows to draw several interesting conclusions on the nature of ε, especially if we change into the complex plane. Provided that $\mathcal{G}$ is finite for all τ we see for

---

[17]Equations (3.25) and (3.37) are examples of the *convolution theorem* of Fourier integrals [5]:

$$c(\omega) = a(\omega)b(\omega) \quad \longleftrightarrow \quad C(t) = \frac{1}{\sqrt{2\pi}} \int\limits_{-\infty}^\infty dt'\, A(t')B(t - t').$$

A convolution in the time domain is translated to a product in the frequency domain and vice versa. The nonlocal connection between D and E therefore is only visible for the time dependent representation.

example that ε is an analytic (holomorphic) function of ω in the upper half-plane (i.e. it is complex differentiable in a neighborhood of every point in this domain). This, in turn, sets the stage for a central and very beautiful statement of complex analysis named after the French mathematician Augustin-Louis Cauchy[18] [40]:

Cauchy's integral formula

$$f(z_0) = \frac{1}{2\pi i} \oint_{C \equiv \partial \Omega} \frac{f(z)}{z - z_0} \, dz. \tag{3.40}$$

If we integrate a complex-valued and analytic function f along the border C of a closed area Ω, we can immediately calculate its value at any point z_0 inside of Ω. Thus the function f is completely determined by its values on the boundary C. If we apply Cauchy's theorem to the susceptibility χ_e we get

$$\frac{\varepsilon(\omega)}{\varepsilon_0} = 1 + \frac{1}{2\pi i} \oint_C \frac{\varepsilon(z)/\varepsilon_0 - 1}{z - \omega} \, dz, \tag{3.41}$$

where the integration contour consists of the real ω axis and a great semicircle with radius $R \to \infty$ in the upper half-plane. If we assume that $\chi_e(\tau)$ does not grow faster than a polynomial, the integration over the semicircle vanishes at infinity and the Cauchy integral reduces to an integration along the real axis, where we have to make an infinitesimal semicircular detour around each pole situated at the axis (e.g. at $z = \omega$). This detour formally introduces a delta function and with the principal value $\mathcal{P}$ we finally get

$$\frac{\varepsilon(\omega)}{\varepsilon_0} = 1 + \frac{1}{i\pi} \mathcal{P} \int_{-\infty}^{\infty} \frac{\varepsilon(\omega')/\varepsilon_0 - 1}{\omega' - \omega} \, d\omega'. \tag{3.42}$$

Taking the real and imaginary part yields the dispersion relations

Kramers-Kronig relations

$$\mathfrak{Re}\left[\frac{\varepsilon(\omega)}{\varepsilon_0}\right] = \frac{1}{\pi} \mathcal{P} \int_{-\infty}^{\infty} \frac{\mathfrak{Im}\left[\varepsilon(\omega')/\varepsilon_0\right]}{\omega' - \omega} \, d\omega' = \frac{2}{\pi} \mathcal{P} \int_{0}^{\infty} \frac{\omega' \mathfrak{Im}\left[\varepsilon(\omega')/\varepsilon_0\right]}{\omega'^2 - \omega^2} \, d\omega',$$

$$\tag{3.43a}$$

---

[18]Born 21st August 1789 in Paris; † 23rd May 1857 in Sceaux.

$$\Im m\left[\frac{\varepsilon(\omega)}{\varepsilon_0}\right]=-\frac{1}{\pi}\,\mathcal{P}\int\limits_{-\infty}^{\infty}\frac{\Re e\left[\varepsilon(\omega')/\varepsilon_0-1\right]}{\omega'-\omega}\,d\omega'=-\frac{2\omega}{\pi}\,\mathcal{P}\int\limits_{0}^{\infty}\frac{\Re e\left[\varepsilon(\omega')/\varepsilon_0-1\right]}{\omega'^2-\omega^2}\,d\omega'.$$

$$(3.43b)$$

Each pole of $\frac{\chi_e(z)}{z-\omega}$ contributes to a certain residue, e.g. for a conductor χ_e has an additional simple pole at $z=0$ and the corresponding residue needs to be added to the equations. With the residue theorem we can rewrite our calculations as

$$\oint_C\frac{\chi_e(z)}{z-\omega}\,dz=-i\pi\sum_k\text{Res}(z_k)+\mathcal{P}\int\limits_{-\infty}^{\infty}\frac{\chi_e(z)}{z-\omega}\,dz,\qquad(3.44)$$

where z_k accounts for the different poles of the integrand and the minus sign before the residue comes from the counter-clockwise circular direction.

We have deduced these very general connections between the real and imaginary part of the dielectric function simply form the causality relation (3.37) and they show another of the major consequences of a frequency dependent dielectric function: The real part of ε is always connected with a corresponding imaginary part and thus a frequency dependent ε is always linked to losses in the medium.

3.3.1 Kramers-Kronig Relations for the Drude Dielectric Function

Let us test the entanglement of the real and imaginary part of the dielectric response for the Drude function (2.4). For the sake of convenience we set $\varepsilon_\infty=1$ and use

$$\frac{\varepsilon_d(\omega)}{\varepsilon_0}=1-\frac{\omega_p^2}{\omega^2+i\gamma_d\,\omega},$$

which yields the corresponding susceptibility

$$\chi_d(\omega)=\frac{\varepsilon_d(\omega)}{\varepsilon_0}-1=-\frac{\omega_p^2}{\omega^2+i\gamma_d\,\omega}.\qquad(3.45)$$

Splitting ε_d into its real and imaginary part gives

$$\Re\left\{\frac{\varepsilon_d(\omega)}{\varepsilon_0}\right\} = \varepsilon_1(\omega) = 1 - \frac{\omega_p^2}{\omega^2 + \gamma_d^2}, \tag{3.46a}$$

$$\Im\left\{\frac{\varepsilon_d(\omega)}{\varepsilon_0}\right\} = \varepsilon_2(\omega) = \frac{\gamma_d \omega_p^2}{\omega(\omega^2 + \gamma_d^2)}. \tag{3.46b}$$

Again we change into the complex plane and the poles of $\frac{\chi_d(z)}{z-\omega}$ now occur at $z = 0$, $z = -i\gamma_d$ and $z = \omega$, thus the associated residues yield

$$\text{Res}(z=0) = \lim_{z \to 0} z \frac{\chi_d(z)}{z - \omega} = \lim_{z \to 0} \frac{-\omega_p^2}{(z-\omega)(z+i\gamma_d)} = -i\frac{\omega_p^2}{\omega\gamma_d}$$

$$\text{Res}(z=-i\gamma_d) = \lim_{z \to -i\gamma_d} (z+i\gamma_d)\frac{\chi_d(z)}{z-\omega} = \frac{\omega_p^2}{\gamma_d^2 - i\gamma_d\omega} = \frac{\omega_p^2}{\omega^2+\gamma_d^2} + i\frac{\omega\,\omega_p^2}{\gamma_d(\omega^2+\gamma_d^2)}$$

$$\text{Res}(z=\omega) = \chi_d(\omega)$$

Hence within the Drude model we finally get

$$\chi_d(\omega) = -\frac{\omega_p^2}{\omega^2+\gamma_d^2} + i\frac{\gamma_d\,\omega_p^2}{\omega(\omega^2+\gamma_d^2)} + \frac{1}{i\pi}\mathcal{P}\int_{-\infty}^{\infty}\frac{\chi_d(\omega')}{\omega'-\omega}\,d\omega'. \tag{3.47}$$

The Krames-Kronig relations with the included new poles now read

$$\varepsilon_1(\omega) - 1 = -\frac{\omega_p^2}{\omega^2+\gamma_d^2} + \frac{1}{\pi}\mathcal{P}\int_{-\infty}^{\infty}\frac{\varepsilon_2(\omega')}{\omega'-\omega}\,d\omega', \tag{3.48a}$$

$$\varepsilon_2(\omega) = \frac{\gamma_d\,\omega_p^2}{\omega(\omega^2+\gamma_d^2)} - \frac{1}{\pi}\mathcal{P}\int_{-\infty}^{\infty}\frac{\varepsilon_1(\omega')}{\omega'-\omega}\,d\omega'. \tag{3.48b}$$

If we alternately insert Eq. (3.46) into the above expressions the principal value integrals vanish in both cases and we immediately recover the previous result for the Drude dielectric function.

3.4 Rayleigh Scattering: The Quasistatic Approximation

Now we have all the tools at hand to finally solve Maxwell's equations for metallic nanoparticles. Before we discuss the solution of the full wave equations in the next section, let us elucidate the main argument for a more simplified case. In the limit of small particles (compared to the wavelength λ, e.g. particles $< 50\,\text{nm}$) we can put $k \approx 0$ and neglect all retardation effects. The wave equation (3.34a) for the scalar potential ϕ and a non vanishing external charge distribution ρ transforms to the Poisson equation

$$\nabla^2 \phi(r) = -\frac{\rho(r)}{\varepsilon} \qquad (3.49)$$

If there are again no external charges present, this equation reduces to the Laplace equation

$$\nabla^2 \phi(r) = 0 . \qquad (3.50)$$

As a result to a work [41] now almost 200 years old, by the extraordinary autodidact George Green,[19] we know that we can solve this kind of differential equation through the introduction of the Green function G

$$\nabla^2 G(r - r') = -4\pi\delta(r - r') . \qquad (3.51)$$

With the Dirac delta distribution on the right hand side, we already obtain a first hint about the singularity of G. In the quasistatic regime the solution of Eq. (3.51) is

Quasistatic Green function

$$G(r - r') = \frac{1}{|r - r'|} = G(r, r') \qquad (3.52)$$

plus an arbitrary function which has to obey the Laplace equation. With this static Green function G we are now prepared to solve the Poisson equation within an unbounded region, simply by applying Green's theorem together with appropriate Dirichlet or Neumann boundary conditions, see [5] or [18] for more details. If we multiply Eq. (3.51) with the right hand side of the general Poisson equation

---

[19]Born 14th July 1793 in Sneinton, Nottingham; † 31st May 1841 in Nottingham. See e.g. [42] or [43] for more details on his life.

$\mathbf{V}^2\phi(r) = f(r)$ and integrate over the region Ω we get

$$\int_\Omega \mathbf{V}^2 G(r,r')f(r')\,\mathrm{d}V' = -4\pi \int_\Omega \delta(r - r')f(r')\,\mathrm{d}V' = -4\pi f(r) = -4\pi \mathbf{V}^2\phi(r),$$

$$\mathbf{V}^2\phi(r) = -\frac{1}{4\pi}\int_\Omega \mathbf{V}^2 G(r,r')f(r')\,\mathrm{d}V'.$$

Because the Laplace operator $\mathbf{V}^2$ (sometimes also denoted with the symbol $\triangle$) is linear and acts only on r, we can interchange derivative and integral and finally write down the solution for Eq. (3.49) as

$$\phi(r) = \frac{1}{4\pi}\int_\Omega G(r - r')\frac{\rho(r')}{\varepsilon}\,\mathrm{d}V'. \tag{3.53}$$

Let us assume a given external excitation ϕ_{ext}, which, for example, corresponds to plane waves impinging on a nanoparticle. The only possible sources are bound to the nanoparticle surface $\partial\Omega$, therefore the charge density ϱ reduces to a surface charge density $\bar{\sigma}$. The actual surface charge of a nanoparticle[20] can always be determined through the boundary condition Eq. (3.10) as

Surface charge of a metallic nanoparticle

$$\int_{\partial\Omega} \hat{n}\cdot(\boldsymbol{D}_2 - \boldsymbol{D}_1)\,\mathrm{d}a = -\int_{\partial\Omega}(\varepsilon_2\phi_2' - \varepsilon_1\phi_1')\,\mathrm{d}a. \tag{3.54}$$

To be consistent with the approach presented in [44] (which will be discussed in the next section), we have to define a new kind of surface charge density as

$$\sigma = \frac{\bar{\sigma}}{4\pi\varepsilon}. \tag{3.55}$$

Note that with this redefinition the units also change from $[\bar{\sigma}] = \text{C}/\text{m}^2$ to $[\sigma] = \text{V}/\text{m}$. The scalar potential as a solution of Eq. (3.49) and an external excitation ϕ_{ext} can now be expressed as[21] [44, 45]

---

[20]Remember that within Maxwell's theory we deal with abrupt interfaces and we additionally assume homogeneous media where ε only depends on the frequency ω.

[21]The redefinition of $\bar{\sigma}$ does not change the unit of the potential, since we still follow Eq. (3.53). Therefore the units of the electromagnetic fields or other quantities also remain unchanged.

$$\phi(r) = \int_{\partial\Omega} G(r - s')\sigma(s')\,\mathrm{d}a' + \phi_{\text{ext}}(r), \tag{3.56}$$

with σ being the artificial surface charge density situated at the interface $\partial\Omega$ between the two media. Note that this equation fulfills the Poisson equation everywhere except at the boundaries.

We make an important step here, which highlights the main idea of our approach in the quasistatic regime: For an unbounded region we can write down the solutions of the Poisson or Laplace equation quite easily. For a bounded region the corresponding boundary conditions come into play and *we add artificially a surface charge density that will be chosen such that the boundary constraints are fulfilled.*

We already discussed the electromagnetic boundary conditions in Sect. 3.2.1: The tangential electric field and the normal component of the dielectric displacement have to be continuous at the boundary between two media. The first constraint implies that the scalar potential is continuous $\phi_1|_{\partial\Omega} = \phi_2|_{\partial\Omega}$, which is guaranteed when the surface charge density is the same on each side of the boundary: $\sigma_1 = \sigma_2$.

For the second constraint $\varepsilon_1\phi_1'|_{\partial\Omega} = \varepsilon_2\phi_2'|_{\partial\Omega}$ we have to evaluate the surface derivative of ϕ

$$\lim_{r\to s}\hat{n}\cdot\nabla\phi(r) \equiv \lim_{r\to s}\frac{\partial\phi(r)}{\partial n} = \lim_{r\to s}\left\{\frac{\partial}{\partial n}\int_{\partial\Omega} G(r - s')\sigma(s')\,\mathrm{d}a' + \frac{\partial\phi_{\text{ext}}(r)}{\partial n}\right\}.$$

Because of the singularity of G we have to be careful about the limit $r \to s$ in the integral. Let us consider

$$\lim_{r\to s}\hat{n}\cdot\nabla\int G(r, s')\sigma(s')\,\mathrm{d}a' \tag{3.57}$$

for a coordinate system where $\hat{n} = \hat{e}_z$, $r = (0, 0, z)^T$, and $s' = \rho(\cos\varphi, \sin\varphi, 0)^T$ is given in polar coordinates ρ and φ, see [46]. We compute the boundary integral within a small circle with radius R, within which the surface charge σ can be approximated by a constant. The integral then becomes

$$\lim_{z\to\pm 0}\hat{n}\cdot\int\frac{r - s'}{|r - s'|^3}\,\mathrm{d}a' \to \lim_{z\to\pm 0} 2\pi z\int_0^R \rho\mathrm{d}\rho\,(\rho^2 + z^2)^{-\frac{3}{2}} = \pm 2\pi. \tag{3.58}$$

The positive or negative sign depends on the direction from which we approach the surface, whether from inside or outside of the particle boundary (i.e. positive sign inside the particle, negative sign outside of it). With the abbreviation $F(s, s') = (\hat{n}\cdot\nabla)G(s, s')$ we finally obtain

$$\frac{\partial\phi(s)}{\partial n} = \int_{\partial\Omega} F(s, s')\sigma(s')\,\mathrm{d}a' \pm 2\pi\sigma(s) + \frac{\partial\phi_{\text{ext}}(s)}{\partial n}. \tag{3.59}$$

3.4.1 From Boundary Integrals to Boundary Elements

The calculation of an analytical solution of Eq. (3.59) is only possible for very restricted geometries, for example if $\partial\Omega$ is a sphere boundary. In Chap. 4 different numerical methods and approaches will be discussed, to obtain solutions for more general shapes of $\partial\Omega$. The method that has been used throughout this work is the so-called Boundary Element Method (BEM). In this approach one approximates the surface charge as a discrete number of points located at the centroids of small surface elements (see Sect. 4.4 for details). Equation (3.59) then becomes

$$
\left(\frac{\partial\phi}{\partial n}\right)_i = \sum_j F_{ij}\sigma_j \pm 2\pi\sigma_i + \left(\frac{\partial\phi_{\text{ext}}}{\partial n}\right)_i ,
\tag{3.60}
$$

or in a compact matrix notation

$$
\frac{\partial\phi}{\partial n} = \boldsymbol{F}\sigma \pm 2\pi\sigma + \frac{\partial\phi_{\text{ext}}}{\partial n} ,
\tag{3.61}
$$

where $\boldsymbol{F}$ is the matrix with elements $F(s_i, s_j)$.

So far so good, but we still have to account for Maxwell's boundary conditions. The continuity of the normal component of the dielectric displacement leads to

$$
\varepsilon_2\left(\boldsymbol{F}\sigma + 2\pi\sigma + \frac{\partial\phi_{\text{ext}}}{\partial n}\right) = \varepsilon_1\left(\boldsymbol{F}\sigma - 2\pi\sigma + \frac{\partial\phi_{\text{ext}}}{\partial n}\right)
$$

$$
2\pi\sigma(\varepsilon_2 + \varepsilon_1) + \boldsymbol{F}\sigma(\varepsilon_2 - \varepsilon_1) = -(\varepsilon_2 - \varepsilon_1)\frac{\partial\phi_{\text{ext}}}{\partial n},
$$

where $\varepsilon_{1,2}$ is the (frequency dependent) dielectric function of the corresponding medium, see Fig. 2.3. From the last expression it follows that

$$
\sigma = -\left\{2\pi\frac{\varepsilon_2 + \varepsilon_1}{\varepsilon_2 - \varepsilon_1}\mathbb{1} + \boldsymbol{F}\right\}^{-1}\frac{\partial\phi_{\text{ext}}}{\partial n} .
\tag{3.62}
$$

With the abbreviation $\boldsymbol{\Lambda} = 2\pi\frac{\varepsilon_2 + \varepsilon_1}{\varepsilon_2 - \varepsilon_1}\mathbb{1}$ we obtain our final result

---

Quasistatic surface charge density

$$
\sigma = -(\boldsymbol{\Lambda} + \boldsymbol{F})^{-1}\frac{\partial\phi_{\text{ext}}}{\partial n} .
\tag{3.63}
$$

Note the elegance of this expression, where material parameters (incorporated by the frequency dependent Λ) and structural properties (established by F) of the problem are fully decoupled.

3.4.2 Eigenmode Expansion

To compute the surface charge density σ according to Eq. (3.63) we have to perform a matrix inversion. The size of the matrix (i.e. the refinement of our surface discretization) is the limiting parameter for the required computation time. In the quasistatic regime we can significantly speed up the simulation if we expand Eq. (3.63) to *plasmon eigenmodes* and reduce the inversion of the fully populated matrix to a much faster inversion of a diagonal matrix. Additionally, it turns out that we do not require a complete set of eigenmodes to obtain very accurate results, see [47].

At first we define left and right eigenvectors σ_k^L and σ_k^R of the surface derivative of the Green function through [48, 49]

$$\langle F, \sigma_k^R \rangle = \lambda_k \sigma_k^R, \quad \langle \sigma_k^L, F \rangle = \lambda_k \sigma_k^L, \tag{3.64}$$

which form a biorthogonal set with $\langle \sigma_k^L, \sigma_{k'}^R \rangle = \langle \sigma_k^R, \sigma_{k'}^L \rangle = \delta_{kk'}$. The inner product is defined through a surface integration:

$$\langle A, B \rangle = \int_{\partial \Omega} A(s) B(s) \, \mathrm{d}s. \tag{3.65}$$

Note that F is a non-Hermitian matrix, but with real eigenvalues, consult [48, 50] for further details.

The functions σ_k^R can be interpreted as the *surface plasmon eigenmodes*, and the response to any external perturbation can be decomposed into these modes viz.

$$\langle \sigma_k^L, (\Lambda + F) \rangle \sigma = (\Lambda + \lambda_k) \sigma_k^L \sigma = -\left\langle \sigma_k^L, \frac{\partial \phi_{\text{ext}}}{\partial n} \right\rangle, \tag{3.66}$$

which leads to

Quasistatic surface charge eigenmode expansion

$$\sigma = -\sum_k \frac{\sigma_k}{\Lambda(\omega) + \lambda_k} \left\langle \sigma_k^L, \frac{\partial \phi_{\text{ext}}}{\partial n} \right\rangle. \tag{3.67}$$

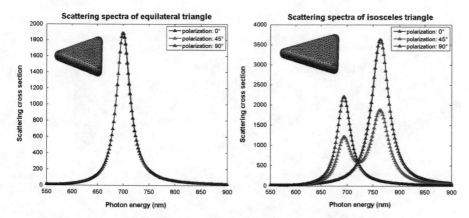

Fig. 3.4 Scattering spectra of an equilateral (*left hand side*) and an isosceles (*right hand side*) gold nanotriangle with three different polarization angles. The basis edge length of each triangle is 50 nm and the isosceles triangle has an altitude of 55 nm. The dielectric surrounding is modeled with an effective refraction index $n_b = 1.33$

It is apparent that a given mode k gives a noticeable contribution only if the coupling $\langle \sigma_k^L, \frac{\partial \phi_{\mathrm{ext}}}{\partial n} \rangle$ to the external potential is sufficiently strong and if the denominator becomes small–this second requirement brings us directly to the plasmon resonance condition that connects our calculated eigenenergy λ_k with the actual photonic energy ω:

$$\mathfrak{Re}\left[\Lambda(\omega) + \lambda_k\right] = 0 . \tag{3.68}$$

We have to be careful here, because Λ is a complex quantity and we assume that the spectral variation of the imaginary part is sufficiently small. For gold particles, this is only true if ω is far away from the d-band absorption. For small gold nanospheres Eq. (3.68) will therefore not give the proper resonance frequency!

Nevertheless the eigenmode expansion is very useful for the physical interpretation of optical phenomena of nanoparticles and allows new insight into the behavior of plasmons. A very revealing illustration of the convenience of eigenmodes is the optical spectrum of an equilateral triangle (also see [51]). If we illuminate such a triangle with polarized light, at first it is a little bit surprising that we do not find any polarization dependence at all in contrast to spectra of other triangles, see Fig. 3.4. Plotting the surface charge at the resonance energy of 700 nm of the equilateral triangle in Fig. 3.4 gives us no hint about its polarization independence, instead we see the expected behavior plotted in Fig. 3.5.

The calculation of the eigenmodes on the other hand shows us very quickly that for an equilateral triangle several eigenmodes are degenerate, whereas (because of the broken symmetry) there is no degeneracy for an isosceles triangle, see Figs. 3.6 and 3.8.

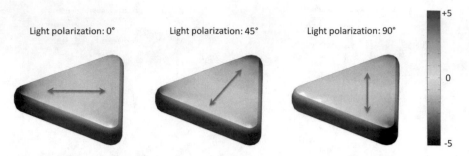

Fig. 3.5 Surface charge in arbitrary units at the resonance energy of 688.8 nm for an equilateral triangle with 50 nm edge length and a height of 10 nm. The particle is illuminated from the top

Fig. 3.6 The first two (degenerate) dipolar eigenmodes of an equilateral triangle with edge length 50 nm and a height of 10 nm

Thus, for equilateral triangles the surface charge distribution in Fig. 3.5 can be generated through a linear combination of the first two optical active and degenerated eigenmodes plotted in Fig. 3.6, yielding always the same peak position independent of the polarization angle. In Fig. 3.7 we show an example for the surface charge distribution with polarization along the x-axis.

The corresponding coefficients of the linear combination can be calculated from $c_k = \langle \sigma_k^L, \sigma \rangle$ and show again that the first two eigenmodes are dominating the surface charge distribution.

3.4.2.1 Identifying Dark Modes with a Hammer

Light usually only couples to those plasmon modes of a nanoparticle, which have a net dipole moment, thus a separation into bright (i.e. radiating) and dark (optically forbidden) modes is common. However, if you want to identify all eigenmodes of your nanosystem, high energetic electrons are a much more suitable excitation than plane light waves. In [52], for example, it has been shown that even for a well studied and simple system like a silver nanodisk there are still new modes to discover which do not couple to light and have thus eluded observation in optical experiments so far. In analogy to corresponding molecular vibrations the new mode is called *dark breathing mode*, since it oscillates with radial symmetry and has no net dipole moment (the dispersion relations follows that of a confined film mode, see [53]). EELS will be discussed later on in Sect. 5.2.2, but a rough and simplified image for the investigation of plasmon eigenmodes with high energetic electrons is given if you think of hitting a drum with a stick or a church bell with a hammer. If you start to swing the bell with the right frequency you will obtain a resonance similar as for plasmons excited by plane waves.[22] If you hit the resting bell with a hammer on the other hand, it will vibrate with the corresponding eigenmodes. In this sense an impinging high energetic electron beam corresponds to hitting a nanoparticle at a very localized spot.[23]

Once the dark breathing mode has been observed with EELS, we were also able to show that retardation effects induced by oblique optical illumination relax

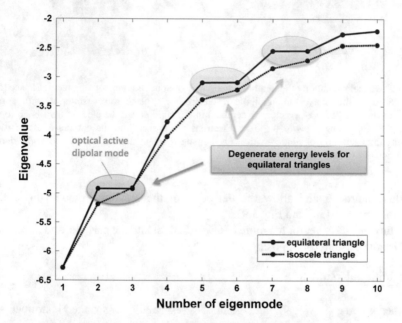

Fig. 3.7 The eigenenergies λ_k of the first ten eigenmodes for an for an equilateral and isosceles triangle with horizontal light polarization, see Eqs. (3.67) and (3.68)

Fig. 3.8 The linear combination of the first two (degenerate) dipolar eigenmodes of an equilateral triangle generates the surface charge distribution plotted in Fig. 3.5

---

[22] By the way, as for all forced oscillations you will require a phase shift of 90° between excitation and the swinging to build up a resonance (also see Fig. 2.9).

[23] Essentially the interaction with the electron beam is a much more complicated process (see Sect. 5.2.2), but the analogy is handy after all.

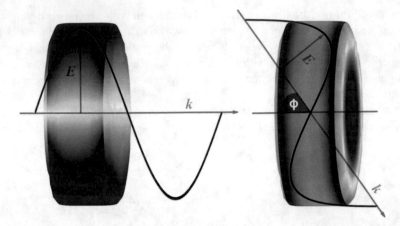

Fig. 3.9 Optical excitation of a breathing mode with oblique illumination of a nanodisk, see [54]. The *left panel* shows a common dipole resonance for a plane wave excitation with normal incidence. For a tilted wave vector, as depicted in the *right panel*, the electric field is no longer constant for opposing sides of the disk. Electrons are pushed into the corners of the particle, allowing the excitation of modes with no net dipole moment. Figure reproduced with friendly permission of Michael Reisecker

the dark character and allow the detection of the breathing mode with optical spectroscopy, see [54] and Fig. 3.9.

Tilting the wave vector fits more of a wavelength into the particle, and the electric field is not constant for opposing sides of the disc.

Additional remark

When it comes to visualizing eigenmodes (especially for stringed instruments like guitars, violins or cellos, see e.g. [55]), the so-called *Chladni figures* are another noteworthy discovery. Over 200 years ago the German physicist Ernst Florens Friedrich Chladni[24] showed how the various vibration modes of a metal plate can be observed by drawing a violin bow along the edge of the plate covered with sand [56]. Once the plate vibration reaches resonance, the sand concentrates along the nodal lines where no oscillation occurs and forms the Chladni patterns.

Nowadays usually a loudspeaker is placed under the plate or membrane to achieve a more accurate vibration, but the Chladni figures remain quite useful and, on a totally different length scale, their equivalent may even serve as distinguishing feature for film and edge modes of particle plasmons, see

---

[24]Born 30th November 1756 in Wittenberg, Electorate of Saxony; † 3rd April 1827 in Breslau, Prussia.

Additional remark

Fig. 3.10 and discussion in [52, 53].

Fig. 3.10 Edge and film eigenmodes with radial and circular node lines, respectively, of a silver nanodisk with a diameter of 200 and 30 nm thickness. The dipole, quadrupole, breathing, and a hybrid mode are shown from *left* to *right*. Figure reproduced with friendly permission of Michael Reisecker

3.5 Solving the Full Maxwell Equations

In the last section we made the assumption that the characteristic length of the investigated system is much smaller than the wavelength of light. For bigger particles or structures this approximation becomes questionable and we have to solve Maxwell equations in their full glory.

Let us first note the main impacts of $k \neq 0$: The vector potential A does not vanish anymore and the electromagnetic fields remain in the form of Eq. (3.32)

$$\mathbf{E} = i\omega\mathbf{A} - \nabla\phi, \qquad \mathbf{B} = \nabla \times \mathbf{A}.$$

As already discussed in Sect. 3.2 we will apply the Lorenz gauge Eq. (3.33)

$$\nabla \cdot A - i\omega\mu\varepsilon\phi = 0.$$

This allows us in principle to consider only the vector potential[25] and to express the scalar one through Eq. (3.33). Nevertheless, in the following we keep both potentials and exploit a scheme where we only require the Green function and its surface derivative (in the interest of readability we use the same symbol G as in the quasistatic approach). The *retarded* Green function now has to obey

$$\left[\nabla^2 + k_j^2\right] G_j(r) = -4\pi\delta(r). \tag{3.69}$$

The solution to this equation is given by

Retarded Green function and surface derivative

$$G_j = \frac{e^{ik_j|r-r'|}}{|r-r'|}, \qquad F_j = \hat{n}_s \cdot \nabla_s G_j(|s - s'|), \tag{3.70}$$

where the subscript j indicates the medium. Accordingly the wavenumber k_j follows as

$$k_j = \omega\sqrt{\mu_j\varepsilon_j}, \tag{3.71}$$

---

[25]It is also possible to establish a numerical approach based on the electromagnetic fields instead of the potentials. But in contrast to our simple collocation scheme, see Sect. 4.4, this usually requires more complex numerical implementations. In the potential-based BEM approach we have to invert matrices of the order $N \times N$, whereas in the field-based BEM approach the matrices are of the order $3N \times 3N$.

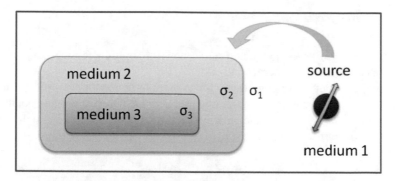

Fig. 3.11 A source in medium 1 can only excite the outer boundary of medium 2, but neither the inner boundary of medium 2 nor the boundaries of medium 3. Also $\sigma_1 \neq \sigma_2 \neq \sigma_3$

where the square root is understood to yield positive imaginary parts (this sign choice ensures that the potentials vanish at infinity, while at the same time, it is consistent with the retarded response formalism) [44]. In general the Green function of a wave equation is the solution of the wave equation for a point source and once it is known, the solution for a general source can be obtained by the principle of superposition [57].

Armed with G and its surface derivative F we are now well prepared to solve Eq. (3.34) inside each medium j analogous to the quasistatic case:

$$\phi(r) = \int_{\partial \Omega_j} G_j(r - s')\sigma_j(s')\, \mathrm{d}a' + \phi_{\text{ext}}(r), \qquad (3.72a)$$

$$A(r) = \int_{\partial \Omega_j} G_j(r - s')h_j(s')\, \mathrm{d}a' + A_{\text{ext}}(r), \qquad (3.72b)$$

where a corresponding redefinition of the surface current density $h = \frac{\mu}{4\pi}\bar{h}$ was necessary ($[h] = \mathrm{Vs/m^2}$). Up to now everything has been very similar to the quasistatic approach and the solutions above automatically fulfill the Helmholtz equations within the different media again. But there are two major differences compared to the quasistatic regime: Identical surface charge and current densities at different boundary sides are in general no longer possible (see discussion in [44]) and we have to choose the external potentials inside the different regions j such that they are only induced by sources neighboring the particle boundaries directly (and are sitting on the correct boundary side), see Fig. 3.11.

3.5.1 Boundary Conditions

Similar to Eq. (3.56) we once again wrote down a solution for the potentials in terms of auxiliary surface charges and currents [Eq. (3.72)], which can be calculated from Maxwell's boundary conditions. Since the expressions are a little bit more laborious than in the quasistatic case, it is useful to introduce shorthand notations. By following [47] we will use the notation hereafter for the scalar potentials inside and outside the particle boundaries:

$$\phi_1 = G_{11}\sigma_1 + G_{12}\sigma_2 + \phi_1^{\text{ext}}, \tag{3.73a}$$

$$\phi_2 = G_{21}\sigma_1 + G_{22}\sigma_2 + \phi_2^{\text{ext}}. \tag{3.73b}$$

The following expressions Eqs. (3.74)–(3.77) are the central part of the fully retarded BEM approach. Let us now exploit the boundary conditions discussed in Sect. 3.2.1. The continuity of the potential can be expressed as

$$G_1\sigma_1 = G_2\sigma_2 + \Delta\phi_{\text{ext}}, \qquad G_1 h_1 = G_2 h_2 + \Delta A_{\text{ext}}, \tag{3.74}$$

$$\sigma_1 = G_1^{-1}(G_2\sigma_2 + \Delta\phi_{\text{ext}}), \qquad h_1 = G_1^{-1}(G_2 h_2 + \Delta A_{\text{ext}}), \tag{3.75}$$

where we have used

$$G_1 = G_{11} - G_{21}, \qquad \Delta\phi_{\text{ext}} = \phi_2^{\text{ext}} - \phi_1^{\text{ext}},$$

$$G_2 = G_{22} - G_{12}, \qquad \Delta A_{\text{ext}} = A_2^{\text{ext}} - A_1^{\text{ext}}.$$

Note that Eq. (3.75) shows us one of the differences to the quasistatic expression mentioned above–the surface charge (and current) on the inside and outside is not identical but rather related through this equation.

3.5.2 Surface Charge and Current Densities

In the next step we have to account for the boundary conditions. Since the calculation is somewhat bulky, we refer to Appendix A.3 and present the surface charge and current densities for the fully retarded regime without further ado:

Retarded surface charge and current densities

$$\sigma_2 = G_2^{-1}\Sigma^{-1}\left[D_e + i\omega\hat{n}(L_1 - L_2)\Delta^{-1}\vec{\alpha}\right], \quad \sigma_1 = G_1^{-1}(G_2\sigma_2 + \Delta\phi_{\text{ext}}), \tag{3.76}$$

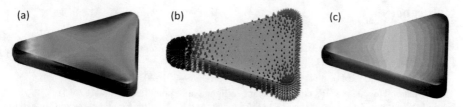

Fig. 3.12 (a) Surface charge density σ_2 of a gold nanotriangle ($55 \times 50 \times 8\,\text{nm}^3$, background refractive index $n_b = 1.34$) at the resonance energy of 792 nm, also see Fig. 4.9. (b) Surface current density h_2 again at the resonance and (c) surface charge density at 1200 nm

$$h_2 = G_2^{-1}\Delta^{-1}\left[i\omega\hat{n}(L_1 - L_2)G_2\sigma_2 + \vec{\alpha}\right], \qquad h_1 = G_1^{-1}(G_2 h_2 + \Delta A_{\text{ext}}).$$
$$(3.77)$$

The corresponding abbreviations used in Eqs. (3.76) and (3.77) can again be found in Appendix A.3. In Fig. 3.12 the surface charge and current density for a gold nanotriangle is exemplified.

References

1. W. Thirring, *Lehrbuch der Mathematischen Physik – Band 2 Klassische Feldtheorie* (Springer, Wien, New York, 1989). ISBN 978-3211821695
2. S. Schaffer, The laird of physics. Nature **471**(7338), 289–291 (2011).
3. B. Mahon, How Maxwell's equations came to light. Nat. Photonics **9**(1), 2–4 (2015).
4. E.T. Whittaker, *A History of the Theories of Aether and Electricity: From the Age of Descartes to the Close of the Nineteenth Century* (University of California Libraries, La Jolla, 2011 (1910)). ISBN 978-1125241103
5. J.D. Jackson, *Classical Electrodynamics* (Wiley, New York, 1962). ISBN 978-0-471-30932-1
6. C. Cohen-Tannoudji, J. Dupont-Roc, G. Grynberg, *Photons and Atoms – Introduction to Quantum Electrodynamics* (Wiley-VCH, New York, 1997). ISBN 978-0-471-18433-1
7. E. Altewischer, W.P. van Exter, J.P. Woerdman, Plasmon-assisted transmission of entangled photons. Nature **418**, 304–306 (2002).
8. M.S. Tame, K.R. McEnery, S.K. Özdemir, J. Lee, S.A. Maier, M.S. Kim, Quantum plasmonics. Nat Phys. **9**(6), 329–340 (2013).
9. M. Schlosshauer, *Decoherence and the Quantum-To-Classical Transition*. The Frontiers Collection (Springer, Berlin/Heidelberg, 2007). ISBN 978-3-540-35773-5
10. M. Schlosshauer, The quantum-to-classical transition and decoherence. arXiv:1404.2635v1, 1–22 (2014).
11. M. Schlosshauer (ed.), *Elegance and Enigma: The Quantum Interviews*. The Frontiers Collection (Springer, Berlin/Heidelberg, 2011).
12. M.O. Scully, M.S. Zubairy, *Quantum Optics* (Cambridge University Press, Cambridge, 1997). ISBN 978-0521435956
13. P.B. Johnson, R.W. Christy, Optical constants of the noble metals. Phys. Rev. B **6**, 12 (1972).
14. J. Zuloaga, E. Prodan, P. Nordlander, Quantum description of the plasmon resonances of a nanoparticle dimer. Nano Lett. **9**, 887–891 (2009).
15. G.G. Szpiro, *Kepler's Conjecture: How Some of the Greatest Minds in History Helped Solve One of the Oldest Math Problems in the World* (Wiley, New York, 2003). ISBN 978-0471086017
16. P. Ball, In retrospect: on the six-cornered snowflake. Nature **480**, 455 (2011).
17. T.C. Hales, A proof of the Kepler conjecture. Ann. Math. **162**, 1065–1185 (2005).
18. A. Zangwill, *Modern Electrodynamics* (Cambridge University Press, Cambridge, 2012). ISBN 9780521896979
19. S.J. Plimpton, W.E. Lawton, A very accurate test of coulomb's law of force between charges. Phys. Rev. **50**, 1066–1071 (1936).
20. A.S. Goldhaber, M.M. Nieto, Terrestrial and extraterrestrial limits on the photon mass. Rev. Mod. Phys. **43**, 277 (1971).
21. R. Lakes, Experimental limits on the photon mass and cosmic magnetic vector potential. Phys. Rev. Lett. **80**, 1826–1829 (1998).
22. W.O. Schumann, Über die strahlungslosen Eigenschwingungen einer leitenden Kugel die von einer Luftschicht und einer Ionosphärenhülle umgeben ist. Z. Naturforsch. A **7**, 149 (1952)
23. W.O. Schumann, Über die Dämpfung der elektromagnetischen Eigenschwingungen des Systems Erde-Luft-Ionosphäre. Z. Naturforsch. A **7**, 250 (1952)
24. M. Balser, C.A. Wagner, Observations of earth-ionosphere cavity resonances. Nature **188**, 638 (1960).
25. P.V. Bliokh, A.P. Nicholaenko, I.F. Filippov, *Schumann Resonances in the Earth-Ionosphere Cavity*. IEE Electromagnetic Waves Series, vol. 9 (1980) Peter Peregrinus, New York. ISBN 978-0906048337

26. H.J. Christian, R.J. Blakeslee, D.J. Boccippio, W.L. Boeck, D.E. Buechler, K.T. Driscoll, S.J. Goodman, J. M. Hall, W.J. Koshak, D.M. Mach, M.F. Stewart, Global frequency and distribution of lightning as observed from space by the optical transient detector. J. Geophys. Res. **108**, 4005 (2003).

27. P.J. Mohr, B.N. Taylor, D.B. Newell, CODATA recommended values of the fundamental physical constants: 2010. Rev. Mod. Phys. **84**, 1527 (2012).

28. D.H. Kobe, Derivation of Maxwell's equations from the gauge invariance of classical mechanics. Am. J. Phys. **48**(5), 348 (1980).

29. D.H. Kobe, Derivation of Maxwell's equations from the local gauge invariance of quantum mechanics. Am. J. Phys. **46**(4), 342 (1978).

30. E. Kapuścik, Generalized Helmholtz theorem and gauge invariance of classical field theories. Lett. Nuovo Cimento Ser. 2 **42**(6), 263–266 (1985).

31. F.J. Dyson, Feyman's proof of the Maxwell equations. Am. J. Phys. **58**, 209–211 (1990).

32. M. Born, *Die Relativitätstheorie Einsteins*, 7th edn. (Springer, Berlin, 2003). ISBN 978-3540004707

33. L. Novotny, B. Hecht, *Principles of Nano-Optics*, 2nd edn. (Cambridge University Press, Cambridge, 2012). ISBN 978-1107005464

34. M. Born, E. Wolf, *Principles of Optics: Electromagnetic Theory of Propagation, Interference and Diffraction of Light*, 7th expanded edn. (Cambridge University Press, Cambridge/New York, 1999). ISBN 0521642221

35. D.L. Mills, *Nonlinear Optics: Basic Concepts* (Springer, Heidelberg, 1998). ISBN 978-3540541929

36. Y.R. Shen, *The Principles of Nonlinear Optics* (Wiley, New York, 1984). ISBN 978-0-471-43080-3

37. M.I. Stockman, D.J. Bergman, C. Anceau, S. Brasselet, J. Zyss, Enhanced second-harmonic generation by metal surfaces with nanoscale roughness: nanoscale dephasing, depolarization, and correlations. Phys. Rev. Lett. **92**, 057402 (2004).

38. T. Hanke, G. Krauss, D. Träutlein, B. Wild, R. Bratschitsch, A. Leitenstorfer, Efficient nonlinear light emission of single gold optical antennas driven by few-cycle near-infrared pulses. Phys. Rev. Lett. **103**, 257404 (2009).

39. L.V. Lorenz, On the identity of the vibrations of light with electrical currents. Philos. Mag. **34**, 287–301 (1867).

40. C.B. Lang, N. Pucker, *Mathematische Methoden in der Physik* (Spektrum Akademischer Verlag, Heidelberg/Berlin, 2010). ISBN 978-3-8274-1558-5

41. G. Green, *An Essay on the Application of Mathematical Analysis to the theories of Electricity and Magnetism* (G. Green, Nottingham, 1828). Reprinted in J. Math. in three parts, digitized version available by Google Books

42. D.M. Cannell, N.J. Lord, George green, mathematician and physicist 1793–1841. Math. Gaz. **77**(478), 26–51 (1993).

43. L. Challis, The green of green functions. Phys. Today **56**(12), 41–46 (2003).

44. F.J. García de Abajo, A. Howie, Retarded field calculation of electron energy loss in inhomogeneous dielectrics. Phys. Rev. B **65**, 115418 (2002).

45. U. Hohenester, J. Krenn, Surface plasmon resonances of single and coupled metallic nanoparticles: a boundary integral method approach. Phys. Rev. B **72**, 195429 (2005).

46. U. Hohenester, A. Trügler, Interaction of single molecules with metallic nanoparticles. IEEE J. Sel. Top. Quantum Electron. **14**, 1430 (2008).

47. U. Hohenester, A. Trügler, MNPBEM – A Matlab toolbox for the simulation of plasmonic nanoparticles. Comput. Phys. Commun. **183**, 370 (2012).

48. R. Fuchs, Theory of the optical properties of ionic crystal cubes. Phys. Rev. B **11**, 1732 (1975).

49. I.D. Mayergoyz, Z. Zhang, G. Miano, Analysis of dynamics of excitation and dephasing of plasmon resonance modes in nanoparticles. Phys. Rev. Lett. **98**, 147401 (2007).
50. A. Messiah, *Quantum Mechanics* (North-Holland, Amsterdam, 1965). ISBN 978-0486409245
51. F.P. Schmidt, H. Ditlbacher, F. Hofer, J.R. Krenn, U. Hohenester, Morphing a Plasmonic Nanodisk into a Nanotriangle. Nano Lett. **14**(8), 4810–4815 (2014).
52. F.-P. Schmidt, H. Ditlbacher, U. Hohenester, A. Hohenau, F. Hofer, J.R. Krenn, Dark plasmonic breathing modes in silver nanodisks. Nano Lett. **12**(11), 5780–5783 (2012).
53. F.-P. Schmidt, H. Ditlbacher, U. Hohenester, A. Hohenau, F. Hofer, J.R. Krenn, Universal dispersion of surface plasmons in flat nanostructures. Nat. Commun. **5**, 3604 (2014).
54. M.K. Krug, M. Reisecker, A. Hohenau, H. Ditlbacher, A. Trügler, U. Hohenester, J.R. Krenn, Probing plasmonic breathing modes optically. App. Phys. Lett. **105**(17), 171103 (2014).
55. C.M. Hutchins, The acoustics of violin plates. Sci. Am. **245**, 170 (1981).
56. E.F.F. Chladni, *Entdeckungen über die Theorie des Klanges* (Breitkopf und Härtel, Leipzig, 1787).
57. W.C. Chew, *Waves and Fields in Inhomogeneous Media* IEEE Press Series on Electromagnetic Waves (Wiley, New York, 1995). ISBN 0-7803-4749-8

Part II
Simulation

Chapter 4
Modeling the Optical Response of Metallic Nanoparticles

Prediction is very difficult, especially if it's about the future.

NIELS BOHR

A formal solution of rigorous scattering theory for nanoparticles is unfortunately only possible for restricted geometries. There exist analytical solutions for light scattering problems if we expand the electromagnetic potentials and fields to spherical harmonics and limit ourselves to spherical or spheroidal particle shapes. In reality, however, we want to work with arbitrary shaped nanostructures and take advantage of certain structure dependent qualities like the hot spots in the gap regions of bowtie antennas or the magnetic response of split-ring resonators. So in general a more sophisticated numerical method for solving Maxwell's equations is essential and inevitable. Several different techniques are available and some of them will be discussed in the next sections.

A critical comparison of the capabilities of the most popular and efficient approaches can be found in [1, 2], for example, an overview is also presented at the end of this chapter. One approximation that all of these methods have in common is the locality of the dielectric description of the material, i.e. the dielectric function depends only on the frequency of light $\varepsilon(r, r', \omega) \approx \varepsilon(\omega)$ and bodies with abrupt interfaces are assumed. This is a valid assumption as long as the studied structures are sufficiently large and finite size effects can be neglected, see Sect. 4.6.3 for more details.

The *optical theorem*[1] relates the forward scattering amplitude to the total cross section of the scatterer, and thus allows us to determine the optical response of metallic nanoparticles.

---

[1]Consult [3] for a derivation and discussion of the theorem, also see [4].

© Springer International Publishing Switzerland 2016

A. Trügler, *Optical Properties of Metallic Nanoparticles*, Springer Series in Materials Science 232, DOI 10.1007/978-3-319-25074-8_4

4.1 Analytic Solutions

4.1.1 Quasistatic Approximation: Rayleigh Theory

Let us start with the quasistatic regime, a topic that has already been discussed in Sect. 3.4. If the investigated structures have spatial dimensions below say 50 nm, we can connect the macroscopic dielectric function with the microscopic polarizability α. The elastic scattering of light can then be described in terms of *Rayleigh scattering*.[2]

For a microscopic derivation of α consult [5] or [6], for example. Expressing the dipole moment $\boldsymbol{P}$ through the *local* microscopic electric field on one hand, and connecting it with the dielectric function ε [through $\boldsymbol{D} = \varepsilon\boldsymbol{E} = \varepsilon_0(\boldsymbol{E} + \boldsymbol{P})$] on the other hand, leads to the *Clausius-Mossotti relation* for spherical particles [6]:

Clausius-Mossotti relation

$$\alpha = 3V\frac{\varepsilon_r - 1}{\varepsilon_r + 2}, \tag{4.1}$$

where V is the volume and $\varepsilon_r = \varepsilon_1/\varepsilon_2$ is the relative dielectric function between medium 1 at the inside of the nanoparticle and medium 2 at the outside, respectively. The expressions for the scattering and absorption cross section then simply follow as

Cross sections within the quasistatic approximation

$$C_{\text{sca}} = \frac{k^4}{6\pi}|\alpha|^2, \qquad C_{\text{abs}} = k\,\Im m\,\{\alpha\}, \qquad C_{\text{ext}} = C_{\text{sca}} + C_{\text{abs}}. \tag{4.2}$$

The quasistatic polarizability of a nanoparticle can also be expressed as the ratio of the induced dipole moment (surface charge times distance) to the electric field of the excitation, which yields the same cross section. The unit of a cross section is generally an area, here and in the following we therefore obtain $[C_{\text{sca}}] = [C_{\text{abs}}] =$

---

[2]When the light from our sun reaches earth, the electromagnetic waves get scattered, mostly elastically, by the molecules and suspensoids in our atmosphere which causes the diffuse sky radiation (the characteristic Fraunhofer lines provoked by spectral absorption also occur in a blue sky spectrum of course). Such atmospheric particles are usually much smaller than the light wavelength and thus are a typical example of Rayleigh scattering. Since the short-wavelength part of the radiation becomes more strongly scattered than the longer wavelengths, the bluish part dominates and yields the color of our sky.

$[C_{ext}] = m^2$. Since the typical length scales in plasmonics are given by nanometers, it is of course much more convenient to express the cross sections in nm^2.

4.1.2 Mie Theory

The solution of Maxwell's equations for spherical particles (or infinitely long cylinders) is named after the physicist Gustav Mie[3] and the expansion for elliptical particles became known as Gans or Mie-Gans theory,[4] see Sect. 4.1.3. An interesting overview about Mie's theory annotated with historical remarks can be found in [7], which was written in 2008 at the occasion of the centenary anniversary of Mie's original publication [8].

A rigorous derivation of Mie's formal solution can be found in [6, 8–10] for example, we will only present the basic steps here. We are going to calculate the time-harmonic electromagnetic field of a sphere of arbitrary size embedded in a linear, isotropic, homogeneous medium. Because of the spherical symmetry of the problem, the use of spherical harmonic functions and a multipole extension of the fields is clearly an advantage. As shown in Appendix A.4.1, we can introduce vector harmonics $\mathcal{M}$ and $\mathcal{N}$ that satisfy the wave Eq. (3.31) and have all the required properties of an electromagnetic field. The scalar function ψ is called the generating function for these vector harmonics, see Eq. (A.30). With ψ the problem of finding solutions for the electromagnetic fields reduces to the comparatively simpler task of finding solutions to the scalar wave equation, see Eq. (A.34). The symmetry of the investigated problem dictates the choice of generating functions, i.e. in our case ψ is a function of spherical coordinates:

Scalar wave equation in spherical coordinates

$$\left[\frac{1}{r^2} \frac{\partial}{\partial r} \left(r^2 \frac{\partial}{\partial r} \right) + \frac{1}{r^2 \sin\theta} \frac{\partial}{\partial \theta} \left(\sin\theta \frac{\partial}{\partial \theta} \right) + \frac{1}{r^2 \sin^2\theta} \frac{\partial^2}{\partial \varphi^2} + k^2 \right] \psi = 0,$$

(4.3)

where $k = n\frac{\omega}{c}$. This equation can be solved by the usual product ansatz for $\psi(r, \theta, \varphi)$, as discussed Appendix A.4 for example. This leads to three decoupled differential equations and after some algebra [6] we can construct the solution

---

[3]Born 29th September 1868 in Rostock; † 13th February 1957 in Freiburg im Breisgau.

[4]At this point usually a little pedantry sets in and we should not speak of it as a theory but rather call it Mie oder Mie-Gans solution, since it is just a result of Maxwell's equations under certain circumstances. Nevertheless the name "Mie theory" has become established.

of (4.3) as a linear combination of even and odd generating functions [9]:

$$\psi_{lm} = \begin{cases} \cos(m\varphi)P_l^m(\cos\theta)z_l(kr), & \text{even solution} \\ \sin(m\varphi)P_l^m(\cos\theta)z_l(kr), & \text{odd solution} \end{cases} \tag{4.4}$$

where $P_l^m(\cos\theta)$ are the associated Legendre functions of the first kind of degree l and order m. The symbol z_l is a substitute for any of the four spherical Bessel functions j_l, y_l, $h_l^{(1)}$, or $h_l^{(2)}$. Because of the completeness of each individual function on the right hand side of Eq. (4.4), ψ_{lm} may serve as a basis and any quantity that fulfills Eq. (4.3) may be expanded as an infinite series in these generating functions. But since we are interested in solutions of the field equations, it is more convenient to go one step backwards and use the vector spherical harmonics as basis:

$$\mathcal{M}_{lm} = \nabla \times (\mathbf{r}\psi_{lm}), \qquad \mathcal{N}_{lm} = \frac{1}{k}(\nabla \times \mathcal{M}_{lm}). \tag{4.5}$$

If u and v are two solutions of Eq. (4.3), we can derive Maxwell's electromagnetic fields in terms of $\mathcal{M}_u$, $\mathcal{N}_u$, $\mathcal{M}_v$, and $\mathcal{N}_v$ (see [6] for the proof)

$$\mathbf{E} = \mathcal{M}_v - \mathrm{i}\mathcal{N}_u, \tag{4.6a}$$

$$\mathbf{H} = -\frac{k}{\mu\omega}(\mathcal{M}_u + \mathrm{i}\mathcal{N}_v). \tag{4.6b}$$

Before we are able to calculate Mie's solution for the scattering of a plane wave, we still have to express the incident plane wave in the same basis functions. This derivation can again be found in many textbooks and the interested reader may once again be referred to Bohren and Huffman's excellent book [6]–and thereby "*acquire virtue through suffering*", in the words of Bohren and Huffman [6]: "*... this is undoubtedly the result of the unwillingness of a plane wave to wear a guise in which it feels uncomfortable; expanding a plane wave in spherical wave functions is somewhat like trying to force a square peg into a round hole.*"

It can be proven that the following two choices of u and v together with Eq. (4.6) generate an adequate expression for the incident plane wave:

$$u = \mathrm{e}^{-\mathrm{i}\omega t}\cos(\varphi)\sum_{l=1}^{\infty}(-\mathrm{i})^l\frac{2l+1}{l(l+1)}P_l^1(\cos\theta)j_l(k_{\mathrm{out}}r), \tag{4.7a}$$

$$v = \mathrm{e}^{-\mathrm{i}\omega t}\sin(\varphi)\sum_{l=1}^{\infty}(-\mathrm{i})^l\frac{2l+1}{l(l+1)}P_l^1(\cos\theta)j_l(k_{\mathrm{out}}r). \tag{4.7b}$$

The field outside the sphere is then given by a superposition of the incident plane wave plus the scattered wave. If we exploit Maxwell's boundary conditions and the conditions to be satisfied at infinity we find the following general expressions:

Outside Mie solution, scattered wave

$$u = e^{-i\omega t} \cos(\varphi) \sum_{l=1}^{\infty} (i)^l a_l \frac{2l+1}{l(l+1)} P_l^1(\cos\theta) h_l^{(2)}(k_{out}r), \tag{4.8a}$$

$$v = e^{-i\omega t} \sin(\varphi) \sum_{l=1}^{\infty} (i)^l b_l \frac{2l+1}{l(l+1)} P_l^1(\cos\theta) h_l^{(2)}(k_{out}r). \tag{4.8b}$$

Similarly, the field inside the sphere is given by

Inside Mie solution

$$u = e^{-i\omega t} \cos(\varphi) \sum_{l=1}^{\infty} (-i)^l c_l \frac{2l+1}{l(l+1)} P_l^1(\cos\theta) j_l(k_{in}r), \tag{4.9a}$$

$$v = e^{-i\omega t} \sin(\varphi) \sum_{l=1}^{\infty} (-i)^l d_l \frac{2l+1}{l(l+1)} P_l^1(\cos\theta) j_l(k_{in}r). \tag{4.9b}$$

The different appearance of the Bessel functions $h_l^{(2)}$ and j_l is due to the asymptotic behavior of the scattered wave and the finite field at the origin, respectively. The expressions of the undetermined coefficients a_l, b_l, c_l, and d_l follow once more from the boundary conditions. After some algebra (that once again can be found in [6, 9]) we derive the Mie coefficients for the outside field with the relative refractive index $n_r = \frac{n_{in}}{n_{out}}$, the relative susceptibility $\mu_r = \frac{\mu_{in}}{\mu_{out}}$, and the abbreviation $x = k_{out}r$

Mie scattering coefficients

$$a_l = \frac{n_r \psi_l(n_r x) \psi_l'(x) - \mu_r \psi_l(x) \psi_l'(n_r x)}{n_r \psi_l(n_r x) \xi_l'(x) - \mu_r \xi_l(x) \psi_l'(n_r x)}, \tag{4.10a}$$

$$b_l = \frac{\mu_r \psi_l(n_r x) \psi_l'(x) - n_r \psi_l(x) \psi_l'(n_r x)}{\mu_r \psi_l(n_r x) \xi_l'(x) - n_r \xi_l(x) \psi_l'(n_r x)}, \tag{4.10b}$$

where we have introduced the *Riccati-Bessel functions*[5]

Riccati-Bessel functions

$$\psi_l(z) = z j_l(z), \qquad \xi_l(z) = z h_l^{(1)}(z). \qquad (4.11)$$

Note that a_l and b_l vanish as n_r and μ_r approach unity: When the particle disappears, so does the scattered field. Likewise we can derive the Mie coefficients for the inside field:

Mie coefficients for inside field

$$c_l = \frac{\mu_r n_r \psi_l(x) \xi_l'(x) - \mu_r n_r \xi_l(x) \psi_l'(x)}{\mu_r \psi_l(n_r x) \xi_l'(x) - n_r \xi_l(x) \psi_l'(n_r x)}, \qquad (4.12a)$$

$$d_l = \frac{\mu_r n_r \psi_l(x) \xi_l'(x) - \mu_r n_r \xi_l(x) \psi_l'(x)}{n_r \psi_l(n_r x) \xi_l'(x) - \mu_r \xi_l(x) \psi_l'(n_r x)}. \qquad (4.12b)$$

---

[5]We follow the notation of Debye here, otherwise the first Riccati-Bessel function is usually denoted as $S_l(z) = z j_l(z)$.

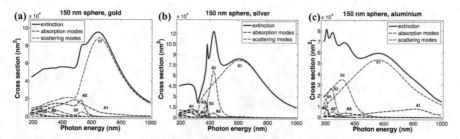

Fig. 4.1 Extinction cross sections of a sphere with 150 nm diameter made of different materials: (**a**) gold, (**b**) silver and (**c**) aluminum [measured values for $\varepsilon(\omega)$ from [11] and the dielectric function database of F. J. García de Abajo]. The spectra can be decomposed into contributions from absorption (A_n) and scattering modes (S_n) with $n = 1, 2, 3, \ldots$ for the dipole, quadrupole, octopole mode, ... respectively (see also [7]). To simplify matters magnetic modes have been omitted here but also give important contributions

4.1.2.1 Cross Sections with Mie Theory

Once the Mie coefficients are determined, we can calculate the extinction, absorption and scattering cross sections (see Fig. 4.1) or the electromagnetic fields inside and outside of the spherical particle.

The cross sections follow from the net rate at which electromagnetic energy crosses the surface of an imaginary and sufficiently large sphere surrounding our particle [6]. They yield

---

Cross sections with Mie theory

$$C_{\text{sca}} = \frac{2\pi}{k_{\text{out}}^2} \sum_{l=1}^{\infty} (2l + 1) \left(\left| a_l^2 \right| + \left| b_l^2 \right| \right), \tag{4.13}$$

$$C_{\text{ext}} = \frac{2\pi}{k_{\text{out}}^2} \sum_{l=1}^{\infty} (2l + 1) \Re\mathrm{e} \left(a_l + b_l \right), \tag{4.14}$$

$$C_{\text{abs}} = C_{\text{ext}} - C_{\text{sca}}. \tag{4.15}$$

A MATLAB® code example for Mie scattering can be found in Appendix B (a dielectric table containing the photon energy in eV and the corresponding real and imaginary part of the refractive index has to be provided, cf. [11]). An analogue example for MATHEMATICA™ can be found in [12] and in [13] a recent test of Mie's predictions for the scattering and absorption of single plasmonic particles has been published.

Additional remark

The optical phenomena in our atmosphere have fascinated their beholders for millennia and one of the most prominent examples are rainbows. They are deeply rooted in arts, culture and mythology, as a symbol of hope and social change, or in the tales of the Irish *leprechaun* or the burning rainbow bridge *Bifröst* in Scandinavian myths. Moreover, after a rain shower, the scattering of sunlight in the remaining water droplets may serve as a beacon of science; if we follow the attempts to explain the origin of rainbows from antiquity to modernity we basically follow the evolution of the theories of light and we meet many brilliant minds from Aristotle and Persian mathematicians [14] through to Descartes and Maxwell along this path. Although the explanation of a rainbow in its full glory remains a hard nut to crack, the basic principles can be quickly explained. With the help of geometrical optics and Snell's law of diffraction, Isaac Newton was one of the first to identify the dispersion of light as one of the main processes behind the colored bows [15], see Fig. 4.2.

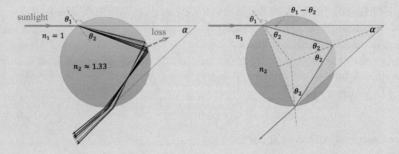

Fig. 4.2 Basic principle of the primary rainbow. The dispersion of light inside a water droplet leads to a splitting into its primary colors, where red light ($n \approx 1.331$) is refracted less than the bluish part ($n \approx 1.343$). A double reflection of the light causes a secondary bow, where the colors become inverted (blue on the outside and red on the inside). With each reflection some of the light intensity is lost.

From the smaller isosceles triangle in the right panel of Fig. 4.2 we obtain for the two angles opposite the equal sides $180 - 2\theta_2$. The sum of the angles in the bigger triangle then yields

$$180 = \alpha + 2(\theta_1 - \theta_2) + (180 - 2\theta_2). \qquad (4.16)$$

With Snell's law $n_1 \sin \theta_1 = n_2 \sin \theta_2$ and $n_1 = 1$ we derive the final scattering angle as

$$\alpha = 4\theta_2 - 2\theta_1 = 4 \arcsin\left(\frac{\sin \theta_1}{n_2}\right) - 2\theta_1. \qquad (4.17)$$

| Additional remark |

If we plot this angle dependence with an averaged $n_2 \approx 1.33$ we will obtain a maximum at $\alpha \approx 42°$, as can be quickly proved by calculating the extremum

$$\frac{\partial \alpha}{\partial \theta_1} = 0 \quad \Longrightarrow \quad \theta_1 = \arccos\left[\sqrt{\frac{1}{3}(n_2^2 - 1)}\right] \quad \Longrightarrow \quad \alpha \approx 42°.$$

(4.18)

In Fig. 4.2 the most simple scenario is plotted, but the light of the sun can also be reflected twice within the droplet yielding a secondary, fainter rainbow appearing approximately 10° outside the primary bow, see Fig. 4.3. Rainbows of order higher than the second are not observed in the atmosphere, they fade into the background illumination [6]. In the laboratory, however, rainbows up to 17th-order have been observed [16]. Between the primary and secondary bow destructive interference of the scattered light leads to a considerably darker region. This dark space is called Alexander's band, named after the Greek philosopher Alexander of Aphrodisias who first described it.

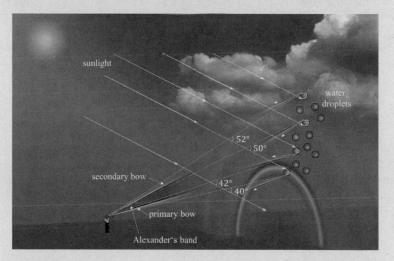

Fig. 4.3 Primary and secondary rainbow caused by the scattering of light in small water droplets. The angular widths of the primary and secondary bow are approximately 1.7° and 3.1°, respectively [6]

The fraction of the total rainbow that can be seen depends on the solar elevation–when the sun is greater than 51° above the horizon, no rainbow can be seen even though conditions are otherwise favorable [6]. From an airplane, on the other hand, it is possible to observe a complete circle rainbow.

The small water droplets in our atmosphere have usual diameters from around 10 μm (e.g. in clouds or fog) up to several mm. For small drops the surface tension of water overtakes the force of air resistance yielding

---

Additional remark

spherical particles. Therefore Mie theory is well suited to calculate the atmospheric scattering of sunlight for such particles and we can derive an analytical solution for the rainbow. But unfortunately the final result still consists of an incredible huge sum of partial waves. In fact, the precise understanding of a rainbow with all its distinct features still remains a hard piece of work (see e.g. [15, 16] for a review). A comprehensive summary about the mathematical physics of rainbows ranging from geometrical optics, the so-called Airy approximation and Mie scattering over complex angular momentum to catastrophe theory, can be found in [17].

In his poem *Lamia* John Keats accused Newton of destroying all the poetry of the rainbow by reducing it to the prismatic colors. The title of Richard Dawkins book *Unweaving the rainbow* [18] is related to this poem and his *"... aim is to guide all who are tempted by a similar view, towards the opposite conclusion."* A similar viewpoint is shown by Richard Feynman writing about the beauty of stars, where he notes that *"... it does not do harm to the mystery to know a little about it. For far more marvelous is the truth than any artists of the past imagined!"*

4.1.3   Mie-Gans Solution

An additional analytical solution for elliptical, spheroidal particles is also possible and is called Mie-Gans solution [6, 9]. Scattering characteristics for oblate and prolate spheroidal particles can be calculated, as long as a quasistatic approximation is adopted.

The three main values α_1, α_2, α_3 of the polarizability tensor can be computed with [9]

Polarizations for Mie-Gans solution

$$\alpha_i = 9V \left(L_i + \frac{1}{\varepsilon_r - 1} \right)^{-1}. \tag{4.19}$$

The geometrical factors L_i are related to the particle shape and always fulfill the sum rule $L_1 + L_2 + L_3 = 1$. Thus for a sphere we obtain $L_i = 1/3$ for all i and we immediately recover the Clausius-Mossotti relation (4.1).

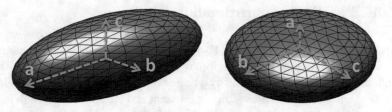

Fig. 4.4 Prolate (*left*, $a > b = c$) and oblate (*right*, $a < b = c$) ellipsoids

For an arbitrary ratio of the particle's semiaxis a, b, and c the factor L_i can be calculated from [9]

$$L_1 = \int\limits_0^\infty \frac{abc\, ds}{2(s + a^2)^{3/2}(s + b^2)^{1/2}(s + c^2)^{1/2}}, \qquad (4.20)$$

with cyclical changes for L_2 and L_3. For the special case of spheroidal particles ($b = c$, see Fig. 4.4), we obtain the solutions [9]

$$\text{prolate } (a > b): \quad L_1 = \frac{1 - e^2}{e^2}\left(-1 + \frac{1}{2e}\ln\frac{1 + e}{1 - e}\right), \quad e = 1 - \left(\frac{b^2}{a^2}\right),$$

$$\text{oblate } (a < b): \quad L_1 = \frac{1 + f^2}{f^2}\left(1 - \frac{1}{f}\arctan f\right), \quad f = \left(\frac{b^2}{a^2}\right) - 1.$$

The corresponding cross sections again follow from Eq. (4.2), with $\alpha = \sum_i \alpha_i \hat{e}_i$, where $\hat{e}$ is the light polarization.

4.2 Discrete Dipole Approximation

The Discrete Dipole Approximation (DDA) is a numerical technique for computing electromagnetic scattering and absorption by targets of arbitrary shape [19], where the continuum target is approximated by a finite point array. In response to the local electric field, each point acquires a dipole moment and the scattering problem can then be solved in a self-consistent way, see Fig. 4.5. Thus, in principle this method is completely flexible regarding the geometry of the target, the only limitation is given by the need to use an interdipole separation that is small compared to any structural lengths in the target and to the wavelength λ [20]. The theoretical basis for the DDA is summarized in [21].

The basic idea of the DDA was already known in 1964, but it was limited to structures that were small compared to the wavelength. This limitation disappeared, when Purcell and Pennypacker introduced the DDA to study interstellar dust grains in 1973 [22]. Fortunately the method is not restricted to astrophysics and Draine and Flatau published a free FORTRAN software package (DDSCAT, current version 7.3) that can be applied to plasmonic scattering problems.

By following [21] very closely (for consistency reasons the Gaussian units from this paper shall also be temporary adopted here), we will shortly show that with the DDA the problem of electromagnetic scattering of an incident light wave can then be cast to the following simple matrix equation

---

Discrete Dipole Approximation

$$\tilde{A} \cdot \tilde{P} = \tilde{E}_{\text{inc}}, \qquad\qquad (4.21)$$

---

where $\tilde{E}_{\text{inc}}$ is a $3N$-dimensional (complex) vector of the incident electric field at the N lattice sites, $\tilde{P}$ is a $3N$-dimensional (complex) vector of the (unknown) dipole polarizations, and $\tilde{A}$ is a $3N \times 3N$ complex matrix.

Let us imagine a point lattice with N occupied sites and an index $j = 1, \ldots, N$ running over these elements. Each dipole j is characterized by a polarizability tensor α_j, which is diagonal with equal components if the material is isotropic (i.e., α_j may be treated as a scalar quantity in this case). We will restrict our attention to instances where all individual dipole polarizability tensors can be simultaneously diagonalized, although it is straightforward to generalize the problem to non diagonal tensors (see [21] for more details). It is nontrivial to choose an adequate α_j for the individual dipoles–Purcell and Pennypacker for example used the Clausius-Mossotti relation (4.1) to obtain an estimate for the polarizability. This assumption is exact in the zero-frequency limit, but it fails at finite ω [21].

Let P_j be the instantaneous complex dipole moment of dipole j, and $E_{\text{ext},j}$ the instantaneous complex electric field at position j due to the incident radiation plus

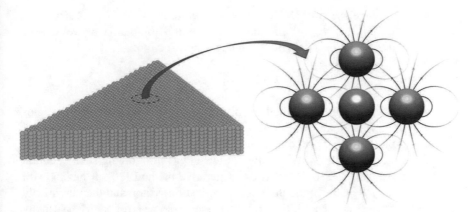

Fig. 4.5 Principle of DDA: the complete volume of a nanoparticle is discretized into a finite set of polarizable points that interact with each other. The total polarization is then calculated in a self-consistent way through an iterative matrix inversion

the other $N-1$ oscillating dipoles. Then we get

$$P_j = \alpha_j E_{\text{ext},j}.$$

(4.22)

As noted in [22], this expression can now be recast as N simultaneous vector equations of the form

$$P_j = \alpha_j \left(E_{\text{inc},j} - \sum_{k \neq j} A_{jk} \cdot P_k \right),$$

(4.23)

where $E_{\text{inc},j}$ corresponds to the electric field of the incident plane wave at position j,

$$E_{\text{inc},j} = E_0 \exp(\mathrm{i}k \cdot r_j - \mathrm{i}\omega t),$$

(4.24)

and $-A_{jk} \cdot P_k$ is the contribution to the electric field at position j due to the dipole at position k. The matrices A_{jk} are defined for $j \neq k$ through

$$A_{jk} \cdot P_k = \frac{e^{\mathrm{i}k r_{jk}}}{r_{jk}^3} \left\{ k^2 r_{jk} \times (r_{jk} \times P_k) + \frac{1 - \mathrm{i}k r_{jk}}{r_{jk}^2} \times \left[r_{jk}^2 P_k - 3 r_{jk}(r_{jk} \cdot P_k) \right] \right\},$$

where $r_{jk} \equiv r_j - r_k$ and $r_{jk} = |r_{jk}|$. By defining the matrix elements for $j = k$ as $A_{jj} = \alpha_j^{-1}$, the scattering problem can be compactly formulated as

$$\sum_{k=1}^{N} A_{jk} \cdot P_k = E_{\text{inc},j} \qquad (j = 1, \ldots, N).$$

(4.25)

If we now introduce the two $3N$-dimensional vectors $\tilde{P} = (P_1, P_2, \ldots, P_N)$ and $\tilde{E}_{\text{inc}} = (E_{\text{inc},1}, E_{\text{inc},2}, \ldots, E_{\text{inc},N})$ and the $3N{\times}3N$ symmetric matrix $\tilde{A}$, we recover the single matrix Eq. (4.21) from the beginning of this section:

$$\tilde{A} \cdot \tilde{P} = \tilde{E}_{\text{inc}}.$$

The most simple (and brute) method to solve this system for the unknown vector $\tilde{P}$ is a direct inversion of $\tilde{A}$. But since $3N$ is a large number (For a typical nanoparticle the number of dipoles is of the order of 10^4 to 10^5!), such a direct method is quite impractical. Many different techniques for solving such equation systems are available and DDSCAT uses an iterative method [20]. It begins with a guess (typically $\tilde{P} = 0$) for the unknown polarization vector, and then iteratively improves the estimate for $\tilde{P}$ until Eq. (4.21) is solved to some error criterion (usually the error tolerance is user-defined and smaller than 10^{-5}, see [20]). Under some circumstances there may be problems with the convergence of this iterative method (see [22], for example), so in general convergence tests are necessary to choose the right grid size.

Once Eq. (4.21) has been solved, the extinction C_{ext}, absorption C_{abs}, and scattering C_{sca} cross sections can be computed from the optical theorem. The results can again be found in [21] and are given by

Cross sections with DDA

$$C_{\text{ext}} = \frac{4\pi k}{|E_{\text{inc}}|^2} \sum_{j=1}^{N} \Im\left(E_{\text{inc},j}^* \cdot P_j\right), \tag{4.26}$$

$$C_{\text{abs}} = \frac{4\pi k}{|E_{\text{inc}}|^2} \sum_{j=1}^{N} \left\{ \Im\left[P_j \cdot (\alpha_j^{-1})^* P_j^*\right] - \frac{2}{3}k^3 P_j \cdot P_j^* \right\}, \tag{4.27}$$

$$C_{\text{sca}} \equiv C_{\text{ext}} - C_{\text{abs}}, \tag{4.28}$$

with the wavenumber $k = 2\pi/\lambda$. When absorption is dominant, it may become difficult to calculate C_{sca} with the above subtraction, because then C_{ext} and C_{abs} have to be computed with high accuracy. It is also possible to compute the scattering cross section directly by calculating the power radiated by the array of oscillating dipoles [21]:

$$C_{\text{sca}} = \frac{k^4}{|E_{inc}|^2} \int d\Omega \left| \sum_{j=1}^{N} \left[P_j - \hat{n}(\hat{n} \cdot P_j)\right] e^{-ik\hat{n}\cdot r_j} \right|^2, \tag{4.29}$$

where $\hat{n}$ is an unit vector in the direction of the scattering, and $d\Omega$ is the element of solid angle.

4.3   Finite Difference Time Domain

The Finite Difference Time Domain (FDTD) method is another volume approach to calculate solutions of Maxwell's equations in a straight forward way. Here the equations in their differential form are discretized in space and time and the time evolution of the electromagnetic near fields is calculated directly (a more elaborate introduction can be found in [23] or [24], for example).

At first, one has to establish the physical region where the fields should be computed, i.e. the nanoparticle structure itself as well as the surrounding volume (see scheme in Figs. 4.6 and 4.7). Usually, absorbing boundary conditions are employed on the edges of this region to ensure that the simulation result is not affected by unwanted back-reflections from these boundaries. The entire computational domain is then discretized and the material property of each grid cell has to be specified. This allows for linear and nonlinear dielectric materials to be implemented quite easily, but a dielectric description in tabulated form, which is the usual outcome of experimental measurements, cannot be used directly. Instead the measured data points have to be approximated by some fit function, again see e.g. [23]. Once the system is implemented, the complete structure can be excited through initial conditions for the fields, be it an exciting localized emitter, an arbitrary incident field or a simple plane wave.

Maxwell's differential equations are then solved directly for every time step, i.e. besides the spatial grid the time variable also has to be discretized. In fact, in the end two shifted temporal and two shifted spatial grids are necessary to compute the desired solution–let's follow the short introduction in [25] and have a look at Faraday's Induction Law Eq. (3.3d) to elucidate this circumstance:

$$\nabla \times \boldsymbol{E}(\boldsymbol{r}, t) = -\frac{\partial \boldsymbol{B}(\boldsymbol{r}, t)}{\partial t}.$$

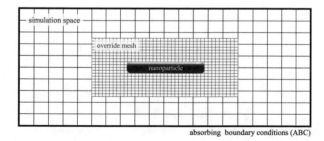

Fig. 4.6 Principle of FDTD: The complete volume of the simulation region is discretized in space and time and at the edges absorbing boundary conditions are implemented. In the region where the scatterer is situated usually a finer override mesh is used. The excitation of the system is realized through initial conditions for the electromagnetic fields, which subsequently are propagated forward in time

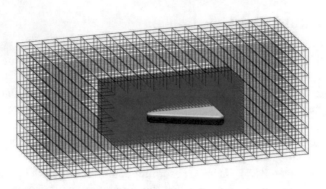

Fig. 4.7 3D view of the FDTD scheme plotted in Fig. 4.6

By approximating the time derivative through a central difference expression

$$\frac{\partial \boldsymbol{B}(\boldsymbol{r}, t)}{\partial t} = \frac{\boldsymbol{B}(\boldsymbol{r}, t + \delta t) - \boldsymbol{B}(\boldsymbol{r}, t - \delta t)}{2(\delta t)}$$

we can rewrite Eq. (3.3d) into

$$\boldsymbol{B}(\boldsymbol{r}, t + \delta t) = \boldsymbol{B}(\boldsymbol{r}, t - \delta t) - 2(\delta t)\nabla \times \boldsymbol{E}(\boldsymbol{r}, t). \tag{4.30}$$

Now the magnetic field $\boldsymbol{B}$ at time $(t + \delta t)$ can be determined if $\boldsymbol{B}$ at time $(t - \delta t)$ and $\boldsymbol{E}$ at time t are known. With a similar procedure we can use Ampère's law (3.3c) to calculate the time propagation for $\boldsymbol{E}$. Consequently the fields $\boldsymbol{E}$ and $\boldsymbol{B}$ are defined on two different time grids, each with a time step of $2\delta t$, but shifted by δt [25]. The spatial discretization follows from the curl operator, which can also be expanded in a sum of central differences. Likewise, in the end we obtain two shifted spatial grids for $\boldsymbol{E}$ and $\boldsymbol{B}$ as well, which are sometimes depicted in a so-called *Yee-cell* [26]. The initial conditions are given by the excitation of the system (e.g. an impinging plane wave) and from there we could start with the forward propagation and directly calculate the temporal evolution of the electromagnetic near-fields. Afterward the *Poynting vector* $S = \boldsymbol{E} \times \boldsymbol{H} = \frac{1}{\mu}\boldsymbol{E} \times \boldsymbol{B}$ gives the directional energy flux and the corresponding far-field spectra can be obtained via some appropriate integral transformations, see e.g. [27].

Additional remark

The formulation of difference quotients is essential for numerical solutions of differential equations and there are different ways to approximate any given differential, see e.g. [28].

Let us start with a differentiable function f and assume that its values can be computed in an area $\pm\delta x$ around x. The second degree Taylor expansion

Additional remark

for $f(x + \delta x)$ and $f(x - \delta x)$ gives

$$f(x + \delta x) = f(x) + f'(x)\delta x + \frac{1}{2!}f''(x)\delta x^2 + \ldots$$

$$f(x - \delta x) = f(x) - f'(x)\delta x + \frac{1}{2!}f''(x)\delta x^2 - \ldots$$

Subtracting these two equations leads to

$$f(x + \delta x) - f(x - \delta x) = 2f'(x)\delta x + \mathcal{O}(\delta x^3)$$

$$f'(x) = \frac{f(x + \delta x) - f(x - \delta x)}{2\delta x} + \mathcal{O}(\delta x^2)$$

For very small δx values, the error of the order of δx^2 can become annoying, so a better approximation with some higher order error term is sometimes quite instrumental. A Taylor expansion for a doubled step size $f(x \pm 2\delta x)$ with a corresponding subtraction of the last formula cancels the expansion term of $\mathcal{O}(\delta x^3)$ and gives the central difference for $\mathcal{O}(\delta x^4)$. With the shorthand notation $f_n = f(x + n\,\delta x)$ we get

$$f'(x) = \frac{-f_2 + 8f_1 - 8f_{-1} + f_{-2}}{12\delta x} + \mathcal{O}(\delta x^4)$$

Analog procedures also allow us to approximate higher order derivatives:

$$f''(x) = \frac{f_1 - 2f_0 + f_{-1}}{\delta x^2} + \mathcal{O}(\delta x^2)$$

$$f''(x) = \frac{-f_2 + 16f_1 - 30f_0 + 16f_{-1} - f_{-2}}{12\delta x^2} + \mathcal{O}(\delta x^4)$$

The expansion of this approach to forward or backward differences, i.e. to problems, where f is known only for values higher or smaller than x respectively, is also straightforward.

4.4 Boundary Element Method

The Boundary Element Method (BEM) has emerged as a powerful numerical technique for solving a wide variety of computational engineering and science problems.[6] The method has deep and complex roots in the history of mechanics[7] and its mathematical foundations include the theorems of Gauss, Green and Stokes; they allow the basic reduction from volume differential equations to boundary integral equations [29] as we have already seen in Chap. 3.

The first step in developing the BEM involves the transformation of the considered set of differential equations [viz. Maxwell's equations, see Eq. (3.34)] into integral equations [we did this in Eqs. (3.56) and (3.72)]. The new integral expressions are valid everywhere–inside and outside the domain as well as on its boundary. Now the structure of the considered problem comes into play: Usually we can solve the equations inside and outside the domain, the only non-trivial part comes from the boundary (note that this approach is based upon a mathematically rigorous definition of the integrals as *limits to the boundary* [29]). As mentioned in Sect. 3.4.1, the BEM is based on an approximation of a continuous surface to a discrete number of points located at the centroids of small surface elements, see Fig. 4.8–this is the so-called collocation method, in contrast to the more complex Galerkin approach [29], for example, where a linear interpolation within the surface elements is used.

Once the surface charges and currents have been derived as discussed in Chap. 3 [see Eqs. (3.63) and (3.76)], we can determine the electromagnetic fields (see Fig. 4.9) and simply compute the scattering, absorption and extinction cross sections again from the Poynting vector [3, 30]:

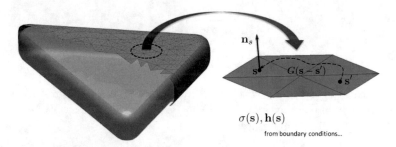

Fig. 4.8 After discretizing the particle surface the corresponding continuous surface charges and currents are approximated as points in the centroids of the small surface elements. The Green function connects these different points with each other, see Eq. (3.70)

---

[6]Examples of application cover the fields of elasticity, geomechanics, structural mechanics, electromagnetics, acoustics, hydraulics, biomechanics, and much more.

[7]A short overview about the historical development of the BEM can be found in [29] for example.

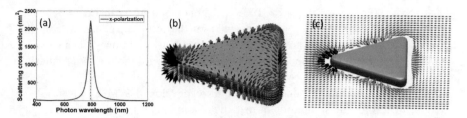

Fig. 4.9 (a) Scattering cross section of a gold nanotriangle ($55 \times 50 \times 8$ nm^3, background refractive index $n_b = 1.34$). The panels (b) and (c) show the electric field at the resonance energy of 792 nm at the particle surface and on the outside, respectively (also see Fig. 3.12)

Cross sections for retarded simulation

$$C_{\text{sca}} = \frac{c}{n_b} \oint_{\partial\Omega} \Re e \left\{ \hat{n} \cdot (E \times B^*) \right\} \, da, \tag{4.31}$$

$$C_{\text{ext}} = -\frac{c}{n_b} \oint_{\partial\Omega} \Re e \left\{ \hat{n} \cdot (E \times B^*_{\text{inc}} + E^*_{\text{inc}} \times B) \right\} \, da, \tag{4.32}$$

$$C_{\text{abs}} = C_{\text{ext}} - C_{\text{sca}}. \tag{4.33}$$

Here we have used that the scattered power can be derived from the time-averaged Poynting vector

$$\langle S \rangle = \frac{1}{2\mu} \Re e \, (E \times B). \tag{4.34}$$

The unit of the Poynting vector is power per area ($[S] = \text{W}/\text{m}^2 = \text{AV}/\text{m}^2$) and we have to integrate the outwardly directed component of the scattered Poynting vector to obtain the scattered power [3]

$$P_{\text{sca}} = \frac{1}{2\mu_b} \oint_{\partial\Omega} \Re e \left\{ \hat{n} \cdot (E \times B^*) \right\} \, da. \tag{4.35}$$

The energy flux density of an incoming plane wave with magnitude E_0 is given by

$$\langle S_0 \rangle = \frac{1}{2} \frac{n_b}{\mu_b c} E_0^2 = \frac{1}{2} \frac{\varepsilon_b c}{n_b} E_0^2 = \frac{1}{2} \sqrt{\frac{\varepsilon_b}{\mu_b}} E_0^2. \tag{4.36}$$

The scattering cross section then finally follows from the radiated scattered power normalized to the incoming energy flux of the plane wave $C_{\text{sca}} = P_{\text{sca}}/\langle S_0 \rangle$. For the sake of convenience we set $E_0 = 1 \, \text{V/m}$ and hence end up with Eq. (4.31). The extinction cross section yields analogously from the total power taken from the incident wave (scattered plus absorbed part) or, as already mentioned, it can also be calculated from the optical theorem, which relates C_{ext} with the forward scattering amplitude [3].

My colleague Ulrich Hohenester and I have developed a MATLAB® toolbox called MNPBEM for the simulation of metallic nanoparticles based on the boundary element method. The toolbox is distributed as free software under the terms of the GNU General Public License, further details can be found in [30, 31]. An update for the inclusion of substrate and layer structure effects has been published in [32].

An open-source Galerkin boundary element library is currently developed at the University College London under the name BEM++ [43–45].

4.5   Other Methods

There exists a huge variety of other methods for the calculation of light scattering and absorption of plasmonic structures. The majority of them are based on a discretization of space (and time). For example some of these approaches are the dyadic Green tensor technique [33, 34], where the dyadic tensor mediates the response between small volume elements of the scatterer, the multiple multipole method [35], the Method of Moments (MoM) [36], or the multiple scattering method [37].

4.6   Comparison Between Different Approaches

In this section we will compare some of the different methods discussed before and highlight the pros and cons of the individual approaches. There exist several papers in the literature where a more extensive comparison is performed, for example see [1, 2] or [37–41] and references therein.

4.6.1   Accuracy

It is possible to generate very accurate results with all the different methods, as long as the discretization is chosen appropriately. In Fig. 4.10 the simulation results of BEM, DDA, and FDTD are compared to the analytical Mie solution.

The agreement between BEM, DDA and FDTD simulations is very good, also if we change to other particle shapes, see Fig. 4.11.

4.6.2   Performance

A performance comparison for the different methods is a challenging task since all of them have their own assets and drawbacks. In this subsection we will only

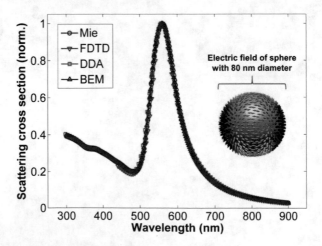

Fig. 4.10   Normalized scattering cross section of a gold sphere with 80 nm diameter in water with $\varepsilon_b = 1.34^2$, calculated for plane wave illumination with x-polarization with different methods. Whereas the BEM result followed almost instantaneously, the DDA simulation took more than one day, see Table 4.2. The FDTD calculation was performed by Johannes Kern (University of Münster)

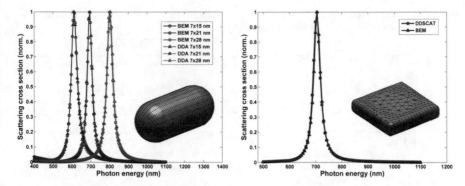

Fig. 4.11 Comparison of the normalized scattering cross section calculated with MNPBEM and DDSCAT for gold rods with changing aspect ratio (*left*) and a $15 \times 15 \times 8\,\text{nm}^3$ gold cube (*right*), $\varepsilon_b = 1.34^2$

Table 4.1 CPU time elapsed for a 10 nm sphere with changing discretization for different kinds of BEM simulations, see [30]

No. of vertices	Nr. of faces	Quasistatic (s)	Eigenmode expansion (s)	Full retarded solution (s)
144	284	2.26	0.23	8.81
256	508	4.83	0.43	32.85
400	796	16.05	0.94	103.96
676	1348	71.82	2.82	446.38

The calculations were performed on a standard office PC running on Windows XP with MAT-LAB® R2013b 32-bit

discuss scattering results for different particle geometries obtained from BEM, DDA and FDTD simulations. For the DDA simulation we used the previously discussed DDSCAT program. The BEM simulations were carried out with the MNPBEM toolbox. The FDTD simulations have been performed by Johannes Kern (University of Münster) with the commercial software LUMERICAL.

In Table 4.1 a computing time comparison for the quasistatic and the full retarded solution of Maxwell's equations as discussed in Chap. 3 is shown. Table 4.2 is adopted in part from [1] with permission of The Royal Society of Chemistry, for details see table caption.

4.6.3 Limits and Inaccuracies

The BEM approach is suited for homogeneous and isotropic dielectric environments, where the embedded bodies are separated by sharp boundaries. Besides the computational limitations, there are some other points that are responsible for inaccuracies or difficult (or impossible) to implement. Nevertheless the BEM works

Table 4.2 Critical comparison of the performance of BEM, DDA, and FDTD adapted from [1] with permission of The Royal Society of Chemistry

	BEM	DDA	FDTD
Computational demand	$V^2 N_\omega$	$V^3 N_\omega$	$V' \omega / \Delta\omega$
Storage demand	$V^{4/3}$	V^2	V'
Average CPU time for 80 nm sphere	1 min	26 h (~2 days)	15 min (4 h)
Advantages	Low computational and storage demand, only surface is discretized	Simple parametrization, only particle volume is discretized	Simple implementation, full spectrum in single run
Disadvantages	Complex parametrization	Time consuming, particle size a and ε limited to $\lvert a\sqrt{\varepsilon}\rvert/\lambda < 5$	Time consuming, difficult to apply to arb. $\varepsilon(\omega)$, parametr. of outside volume and ABCs

N_ω is the number of calculated frequencies and V is the particle volume measured in units of the cube of the skin depth [see Eq. (2.13), $\zeta \approx 15$ nm in the visible near-infrared (see Appendix A.1 for the frequency range of the visible spectrum)]. V' is the discretized volume for the FDTD simulation and it extends considerably outside the particle in order to account for light propagation in the surrounding medium [also absorbing boundary conditions (ABCs) have to be applied, see Sect. 4.3]. In this table for the BEM and DDA approach a direct inversion of the secular linear equations is assumed (DDSCAT uses conjugate gradient algorithms to solve the corresponding equations in an iterative approach–three different algorithms can be chosen in the current version of DDSCAT [20]. DDA scales as $V \log V N_\omega$ when using the iterative method, whereas BEM scales as $V^{4/3} N_\omega$ in this case, see [1]). The factor $\omega/\Delta\omega$ in FDTD is proportional to the number of time steps needed to describe a component of frequency ω with frequency resolution $\Delta\omega$. For the comparison of a typical average computing time the optical spectrum of a gold sphere with a diameter of 80 nm and $N_\omega = 150$ has been calculated (particle symmetry has not been exploited). The DDA simulation ran on 32 nodes of a SUN Fire V20z cluster with $2 \times$ AMD Opteron 248 2 (2 GHz) processors each (the number in brackets corresponds to the simulation time on only 4 nodes to allow a better comparison), whereas the BEM simulation was started on a transtec CALLEO 431L server with four AMD six-core Opteron F 2431 (2.4 GHz) processors and 64 GB DDR2-667 memory. On a standard PC the same BEM simulation takes about 2 min, see Table 4.1

remarkably well and the agreement with experimental data is striking. Here is a short list of possible problems that are inherent to all the discussed simulation methods (fortunately they mostly become important only if the nanoparticles are very small):

- The sharp and abrupt two dimensional Maxwell boundary conditions become questionable for very small particles because of the *spill-out effect* [7]: The surfaces boundary changes to a three dimensional inhomogeneous interface (see Fig. 8.2).
- The surface and interface gets charged because of the different chemical potential of the particle and dielectric background (chemical interface damping [42]).

- The dielectric function of a nanoparticle is different from the dielectric function of bulk material.
- As already mentioned, the dielectric function $\varepsilon(\omega)$ is not homogeneous, it is a local spatial dependent function $\varepsilon(\boldsymbol{r},\boldsymbol{r'},\omega)$ and changes continuously while approaching the surface or defects in the material [7], see Chap. 8.
- The idea of a surface charge density or a surface current density is an idealization of macroscopic electromagnetism. In reality the charge or current is confined to the immediate neighborhood of the surface [3]. (Nevertheless this idealization works quite well and is an essential part of classical field theory.)

References

1. V. Myroshnychenko, J. Rodríguez-Fernández, I. Pastoriza-Santos, A.M. Funston, C. Novo, P. Mulvaney, L.M. Liz-Marzán, F.J.G. de Abajo, Modelling the optical response of gold nanoparticles. Chem. Soc. Rev. **37**, 1792–1805 (2008).
2. W.L. Barnes, Comparing experiment and theory in plasmonics. J. Opt. A Pure Appl. Opt. **11**(11), 114002, (2009).
3. J.D. Jackson, *Classical Electrodynamics* (Wiley, New York, 1962). ISBN 978-0-471-30932-1
4. R.G. Newton, Optical theorem and beyond. Am. J. Phys. **44**, 639–642 (1976).
5. N.W. Ashcroft, N.D. Mermin, *Festkörperphysik* (Oldenbourg, München, 2007). ISBN 978-3-48658273-4
6. C.F. Bohren, D.R. Huffman, *Absorption and Scattering of Light by Small Particles* (Wiley-Interscience, New York, 1983). ISBN 978-0471293408
7. U. Kreibig, Hundert Jahre Mie-Theorie. Phys. Unserer Zeit **39**(6), 281–287 (2008).
8. G. Mie, Beiträge zur Optik trüber Medien, speziell kolloidaler Metalllösungen. Ann. Phys. **330**(3), 377–445 (1908).
9. H.C. van de Hulst, *Light Scattering by Small Particles* (Dover, New York, 1981). ISBN 978-0486642284
10. J.A. Stratton, *Electromagnetic Theory*. (Wiley-IEEE Press, New York, 2007). ISBN 978-0-470-13153-4
11. P.B. Johnson, R.W. Christy, Optical constants of the noble metals. Phys. Rev. B **6**, 12 (1972).
12. C. Sönnichsen, *Plasmons in metal nanostructures*. Ph.D. thesis, Fakultät für Physik der Ludwig-Maximilians-Universität München, 2001.
13. A. Tcherniak, J.W. Ha, S. Dominguez-Medina, L.S. Slaughter, S. Link, Probing a century old prediction one plasmonic particle at a time. Nano Lett. **10**(4), 1398–1404 (2010).
14. J. Al-Khalili, In retrospect: book of optics. Nature **518**(7538), 164–165 (2015).
15. H.M. Nussenzveig, The theory of the rainbow. Sci. Am. **236**, 116–127 (1977).
16. J.D. Walker, Multiple rainbows from single drops of water and other liquids. Am. J. Phys. **44**(5), 421–433 (1976).
17. J.A. Adam, The mathematical physics of rainbows and glories. Phys. Rep. **356**(4–5), 229–366 (2002).
18. R. Dawkins, *Unweaving the Rainbow* (Mariner Books, New York, 2000). ISBN 978-0618056736
19. B.T. Draine, P.J. Flatau, Discrete-dipole approximation for scattering calculations. J. Opt. Soc. Am. A **11**(4), 1491 (1994).
20. B.T. Draine, P.J. Flatau, User guide to the discrete dipole approximation code DDSCAT 7.3. Physics.comp-ph 1–102 (2013), arXiv:1305.6497.
21. B. Draine, The discrete-dipole approximation and its application to interstellar graphite grains. Astrophys. J. **333**, 848–872 (1988).
22. E.M. Purcell, C.R. Pennypacker, Scattering and absorption of light by nonspherical dielectric grains. Astrophys. J. **186**, 705–714 (1973).
23. A. Taflove, *Computational Electrodynamics: The Finite-Difference Time-Domain Method*. Artech House antennas and propagation library, 3rd edn. (Artech House, Boston, 2005). ISBN 1580538320
24. U.S. Inan, R.A. Marshall, *Numerical Electromagnetics – The FDTD Method* (Cambridge University Press, Cambridge, 2011). ISBN 978-0521190695. http://www.cambridge.org/us/academic/subjects/engineering/electromagnetics/numerical-electromagnetics-fdtd-method?format=HB&isbn=9780521190695

25. M. Pelton, G.W. Bryant, *Introduction to Metal-Nanoparticle Plasmonics* (Wiley, Hoboken, 2013). ISBN 9781118060407
26. K. Yee, Numerical solution of initial boundary value problems involving Maxwell's equations in isotropic media. IEEE Trans. Antennas Propag. **14**(3), 302–307 (1966).
27. K. Umashankar, A. Taflove, A Novel Method To Analyze Electromagnetic Scattering Of Complex Objects. IEEE Trans. Electromagn. Compat. **EMC-24**(4), 397–405 (1982).
28. J.H. Mathews, *Numerical Methods Using MATLAB*, 4th edn. (Pearson, Upper Saddle River, 2004). ISBN 0130652482
29. A. Sutradhar, G.H. Paulino, L.J. Gray, *Symmetric Galerkin Boundary Element Method* (Springer, Berlin/Heidelberg, 2008). ISBN 978-3-540-68770-2
30. U. Hohenester, A. Trügler, MNPBEM – A Matlab toolbox for the simulation of plasmonic nanoparticles. Comput. Phys. Commun. **183**, 370 (2012).
31. U. Hohenester, Simulating electron energy loss spectroscopy with the MNPBEM toolbox. Comput. Phys. Commun. **185**(3), 1177–1187 (2014).
32. J. Waxenegger, A. Trügler, U. Hohenester, Plasmonics simulations with the MNPBEM toolbox: consideration of substrates and layer structures. Comput. Phys. Commun. **193**, 138–150 (2015).
33. M. Paulus, P. Gay-Balmaz, O.J.F. Martin, Accurate and efficient computation of the Green's tensor for stratified media. Phys. Rev. E **62**, 5797 (2000). doi:10.1103/PhysRevE.62.5797. http://journals.aps.org/pre/pdf/10.1103/PhysRevE.62.5797
34. P. Gay-Balmaz, O.J.F. Martin, A library for computing the filtered and non-filtered 3D Green's tensor associated with infinite homogeneous space and surfaces. Comput. Phys. Commun. **144**, 111–120 (2002). doi:http://dx.doi.org/10.1016/S0010-4655(01)00471-4. http://ac.els-cdn.com/S0010465501004714/1-s2.0-S0010465501004714-main.pdf?_tid=aab6a980-e476-11e5-9af0-00000aab0f01&acdnat=1457363582_2b3036286ce443a5d0bafda934d16e13
35. L. Novotny, B. Hecht, *Principles of Nano-Optics*, 2nd edn. (Cambridge University Press, Cambridge, 2012). ISBN 978-1107005464
36. W.C. Chew, *Waves and Fields in Inhomogeneous Media*. IEEE PRESS Series on Electromagnetic Waves (IEEE Press, New York, 1995). ISBN 0-7803-4749-8
37. F.J. García de Abajo, Optical excitations in electron microscopy. Rev. Mod. Phys. **82**, 209 (2010).
38. M.A. Yurkin, A.G. Hoekstra, The discrete dipole approximation: an overview and recent developments. J. Quant. Spectrosc. Radiat. Transf. **106**, 558–589 (2007). Also available at arXiv:0704.0038v1.
39. T. Wriedt, U. Comberg, Comparison of computational scattering methods. J. Quant. Spectrosc. Radiat. Transf. **60**(3), 411–423 (1998).
40. J.W. Hovenier, K. Lumme, M.I. Mishchenko, N.V. Voshchinnikov, D.W. Mackowski, J. Rahola, Computations of scattering matrices of four types of non-spherical particles using diverse methods. J. Quant. Spectrosc. Radiat. Transf. **55**(6), 695–705 (1996).
41. M.R. Gonçalves, Plasmonic nanoparticles: fabrication, simulation and experiments. J. Phys. D: Appl. Phys. **47**(21), 213001 (2014).
42. U. Kreibig, Interface-induced dephasing of Mie plasmon polaritons. Appl. Phys. B **93**, 79–89 (2008).
43. W. Śmigaj, S. Arridge, T. Betcke, J. Phillips, M. Schweiger, Solving boundary integral problems with BEM++. ACM Trans. Math. Softw. **41**, 6:1–6:40 (2015).
44. E. van 't Wout, P. Gélat, T. Betcke, S. Arridge, A fast boundary element method for the scattering analysis of high-intensity focused ultrasound. J. Acoust. Soc. Am. **138**, 2726 (2015).
45. S.P. Groth, A.J. Baran, T. Betcke, S. Havemann, W. Śmigaj, The boundary element method for light scattering by ice crystals and its implementation in BEM++. J. Quant. Spectrosc. Radiat. Transf. **167**, 40–52 (2015).

Part III
Implementations and Applications

Chapter 5
Imaging of Surface Plasmons

> *What we observe is not nature itself, but nature exposed to our method of questioning.*
>
> WERNER HEISENBERG

Maxwell's equations are about 150 years old, so why (without any intention to be blasphemous) can we still do interesting physics with them? The answer lies in the interplay with sources: Electromagnetic interactions down at the nanoscale open up the fascinating field of nanooptics, interactions with quantum objects allow us to enter the prosperous world of quantum optics (see Table 5.1).

5.1 Principles of Near-Field Optics

What do we require to produce an optical image of a certain object? Basically there is just one essential thing: A pinhole. One of the most rudimentary imaging systems, the camera obscura, is a very good example of that and has fascinated people throughout the centuries. An optical lens is required to focus more light on the image and in this sense to reduce the exposure time, but the imaging itself comes from diffraction at the aperture. In optical microscopy, the light emitted from an object is diffracted at the boundary of a lens and focused to an image plane by virtue of the refractive index contrast, where usually an intensity profile is detected. If we want to distinguish two small objects close two each other, the resolution depends on the overlap of the two intensity profiles (see Fig. 5.1).

In free space, the propagation of light is determined by the dispersion relation $\hbar\omega = cp = c\hbar k$, which connects the wave vector $k = \sqrt{k_x^2 + k_y^2 + k_z^2} = 2\pi/\lambda$ of a photon (for a real lens the wavelength λ must be corrected by the numerical aperture NA) with its angular frequency ω [1]. Heisenberg's uncertainty principle now already predicts a fundamental resolution limit for optical microscopes: The product of the uncertainty of the momentum of a microscopic particle Δk_x and the

© Springer International Publishing Switzerland 2016
A. Trügler, *Optical Properties of Metallic Nanoparticles*, Springer Series in Materials Science 232, DOI 10.1007/978-3-319-25074-8_5

Table 5.1 Characteristic
length scales L of different
parts of optics compared to
the wavelength λ

Coherent (classical) optics	L	$(\gg) >$	λ
Diffraction grating	L	$>$	λ
Photonic crystals	L	$(\ll) <$	λ
Metamaterials	L	$\ll$	λ

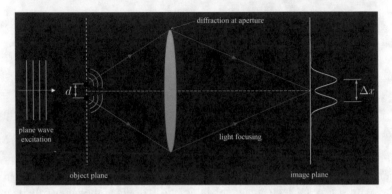

Fig. 5.1 Basic principle of spatial resolution in optical microscopy. Two small sources separated
by a distance d emit light which is collected by a lens far away (compared to the size of the sources)
from the object plane. The numerical aperture NA $= n \sin \theta$ is a characteristic number for the range
of angles over which the lens can collect light. In a good approximation the light wave impinging
at the lens behaves as a plane wave which is diffracted at the boundary of the lens. In the focal or
image plane we can than detect the overlapping intensity profiles

uncertainty in the spatial position Δx in the same direction cannot become smaller
than $\hbar/2$:

$$\hbar \Delta k_x \Delta x \geq \frac{\hbar}{2}. \tag{5.1}$$

Since the maximal possible value of k_x is the total length of the free-space wave
vector $2\pi/\lambda$, we can rewrite the uncertainty equation to

$$\Delta x \geq \frac{1}{2\Delta k_x} = \frac{\lambda}{4\pi}. \tag{5.2}$$

This states that the accomplishable spatial confinement for photons is inversely proportional to the spread[1] in the magnitude of the associated wave vector component
Δk_x.

---

[1] Such a spread in wave vector components occurs for instance in a light field that converges toward
a focus, e.g. behind a lens [1].

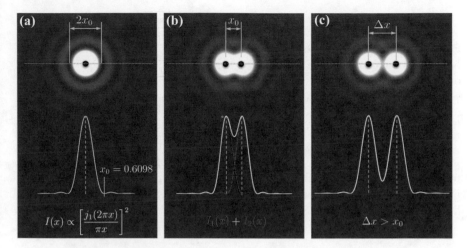

Fig. 5.2 (a) The intensity profile or so-called *Airy disk* of a circular aperture (like the human eye) is proportional to the spherical Bessel function squared divided by πx squared. (b) Resolution limit of two distinguishable maxima defined by the *Rayleigh criterion*, where the second intensity maximum directly falls into the first zero of the first intensity maximum at $x_0 = 0.6098$. If the separation between the two maxima is smaller, the individual peak can no longer be resolved. The value of x_0 also enters Eq. (5.3). (c) If the separation $\Delta x > x_0$, two separate peaks can be detected

The expression (5.2) is very similar to the diffraction-limited resolution derived by Lord Rayleigh [2] (also see Fig. 5.2) or Ernst Abbe[2] [3] in the late nineteenth century:

$$\Delta x = 0.6098 \frac{\lambda}{\mathrm{NA}}. \tag{5.3}$$

We see that there is some arbitrariness in the definition of a resolution limit: We have started with the uncertainty principle of Quantum mechanics, Lord Rayleigh investigated a grating spectrometer [1], while Abbe based his formulation on the distinguishability of the image pattern of two point dipoles. A simple Fourier transformation from position to momentum space also leads us to a similar result:

$$\Delta k_x \Delta x \geq 1, \implies \Delta x \geq \frac{\lambda}{\pi \, \mathrm{NA}} \approx 0.3183 \frac{\lambda}{\mathrm{NA}}, \tag{5.4}$$

see [1] for more details. This last expression is almost two times less pessimistic than Eq. (5.3), but it seems that we cannot obtain a better resolution than approximately

---

[2]Born 23rd January 1840 in Eisenach; † 14th January 1905 in Jena.

half of the wavelength. Does this predict that there is no reasonable chance to investigate objects on the nanoscale without violating fundamental laws of nature? Fortunately the answer is no, there are several ways to elude this limitation. On one hand we can apply some tricks based on the exploitation of distinctive features of the investigated objects. For example we can increase the resolution trough selective excitation of two close-by molecules, if they are distinguishable in energy, orientation or something else.[3]

But on the other hand if we look at Eq. (5.2) once again, we see that in theory there is no limit to optical resolution if the bandwidth of Δk_x is arbitrarily large. In this sense if we include evanescent fields in optical microscopy we are able to increase the resolution limit and to investigate near fields and structures at the nanoscale [5]. We can exploit and detect the high spatial information contained in the evanescent modes (see Fig. 5.3) by probing the electromagnetic field in close proximity of the sources–this is the main principle of near-field nanooptics.

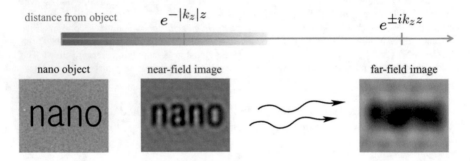

Fig. 5.3 Near-field versus far-field optics: The electric field of an object is given by a Fourier transform and thus connected to the wave vector k. High spatial information about the specimen at the nanoscale is contained in the evanescent modes that decay exponentially with the distance from the object. The conversion to propagating modes which can be detected in the far-field leads to a certain loss of information, since the evanescent contribution is effectively removed from the image

---

[3]The Nobel Prize in chemistry in 2014 was awarded to Eric Betzing, Stefan Hell and William Moerner for the development of super-resolved fluorescence microscopy, a fascinating technique based on the selective deactivation of fluorophores by stimulated emission. A review worth reading about far-field optical nanoscopy can be found in [4]. These methods allowed the direct observation of biological molecules, viruses, cellular protein movements in a living cell, DNA-strands and many other nanoscale objects that would have been destroyed by other microscopes with atomic resolution.

5.2   How to Picture a Plasmon

The previously discussed increase of the resolution limit is neatly realized in scanning near-field optical microscopy (SNOM) experiments for example, where we are able to map the photonic local density of states (LDOS).[4] Another method that allows us to directly picture surface plasmon polaritons of a metallic nanoparticle is Electron Energy Loss Spectroscopy (EELS) [7, 32–35]. In this approach a beam of electrons with high kinetic energy (of the order of 100 keV) passing by (or through) a metallic nanoparticle excites a plasmon and therefore loses a small fraction of its energy. By detecting this energy loss and raster scanning the electron beam across the particle, we obtain a spectral mapping of the plasmon oscillation. We can also combine this technique with EFTEM, where an energy-selecting slit is placed before the detector. Only electrons with a fixed energy range can pass through and we obtain a picture of the plasmon mode at a certain energy. The measurement data can be visualized in a 3D data cube, see Fig. 5.4.

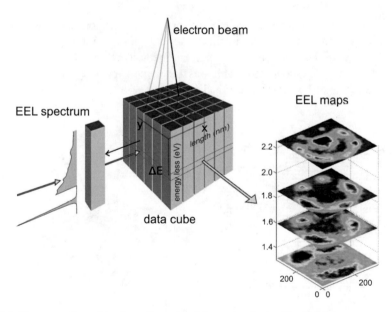

Fig. 5.4 Demonstration of the three-dimensional data cube obtained by EELS measurements. The energy loss spectra are contained in the z-direction (ΔE-axis), whereas the EFTEM maps are stored in the (x, y)-plane. Figure reproduced with friendly permission of Franz Schmidt (University of Graz)

---

[4]The LDOS is a measure of how good a quantum emitter can couple to a nanostructure and it is connected to the imaginary part of the system's dyadic Green function [1, 6]. An increased LDOS through the presence of a nanoparticle allows the emitter to dissipate its energy much more efficient, for example.

5.2.1 Mapping the Plasmonic LDOS

The principal idea for scanning near-field optical microscopy (SNOM) is to bring a detector very close to the surface of the investigated structure–so close that it can detect the evanescent fields. Such an detector can be realized by a fluorophore for example, since its spontaneous emission rate serves as a direct probe of the photonic LDOS at the fluorophore position. In [8] we exploit this capability for mapping the plasmonic modes of gold nanoparticles. We use combined regular arrays of identical gold and fluorophore-doped polymer nanoparticles with a slightly different grating constant. This setup enables the generation of an optical Moiré pattern corresponding to a 200× magnified map of the plasmonic mode, which can be directly imaged with an optical microscope, see Fig. 5.5 and [8].

5.2.2 Electron Energy Loss Spectroscopy

As mentioned above another approach for imaging a plasmon is energy-filtered transmission electron microscopy in the low-energy-loss region of EELS experiments. By varying the position of the high energetic electron beam one can once again map out the evanescent fields. This was first demonstrated by Bosman et al. [9] and Nelayah et al. [10] in 2007, see Fig. 5.6.

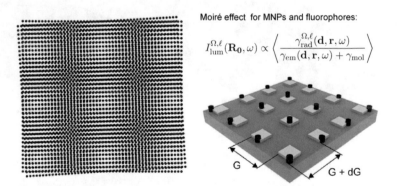

Moiré effect for MNPs and fluorophores:

$$I_{\text{lum}}^{\Omega,\ell}(\mathbf{R_0}, \omega) \propto \left\langle \frac{\gamma_{\text{rad}}^{\Omega,\ell}(\mathbf{d}, \mathbf{r}, \omega)}{\gamma_{\text{em}}(\mathbf{d}, \mathbf{r}, \omega) + \gamma_{\text{mol}}} \right\rangle$$

G

$G + dG$

Fig. 5.5 Principal scheme of optical Moiré effect and the discussed MNP and fluorophore arrays lying upon another. The equation depicts the ensemble average of the luminescence intensity, which is governed by the plasmonic LDOS and describes the probability that an excited molecule emits a photon into the direction of the photodetector. The electromagnetic decay rate is given by $\gamma_{\text{em}}(\mathbf{d}, \mathbf{r}, \omega)$, γ_{mol} is the intra-molecular decay rate and $\gamma_{\text{rad}}^{\Omega,\ell}$ is the differential cross section for light emission into a sphere segment Ω. For a detailed discussion see [8]

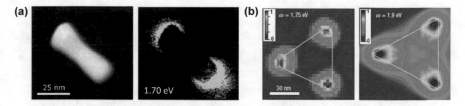

Fig. 5.6 First reported plasmon mapping with EELS experiments. Figure **(a)** reprinted with permission from [9] (© IOP Publishing. Reproduced by permission of IOP Publishing. All rights reserved.), **(b)** reprinted by permission from Macmillan Publishers Ltd: Nature Physics [10], © 2007.

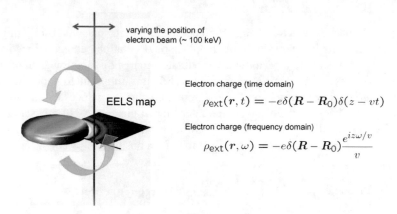

varying the position of
electron beam (~ 100 keV)

EELS map

Electron charge (time domain)

$$\rho_{ext}(r, t) = -e\delta(R - R_0)\delta(z - vt)$$

Electron charge (frequency domain)

$$\rho_{ext}(r, \omega) = -e\delta(R - R_0)\frac{e^{iz\omega/v}}{v}$$

Fig. 5.7 Principle of EELS interaction: the electron beam excites a surface plasmon which in turn acts back on the electrons at a later position. The expression of the electron charge ρ_{ext} corresponds to a charged wire

The dispersion relation of a free electron does not cross the light cone at any point [11], so at first it is not obvious how to couple an electron with light.[5] Besides nonlinear processes or electrons with less momentum than their free counterpart (e.g. from photoemission) it is primarily the interaction with light of higher momentum (evanescent fields) that allows the interaction.

The high-energetic electrons move with 70–80 % of the speed of light and thus relativistic effects come into play. During one plasmon oscillation the swift electrons are able to move about 100 nm–this is sufficient for the plasmon-electron interaction, a process reminiscent of a self energy schematically shown in Fig. 5.7. In a quantum picture, the electron loses energy through the excitation of a surface plasmon. In the

---

[5]An excellent review about optical excitations in electron microscopy can be found in [11] or in [12] with a focus on EELS and cathodoluminescence.

classical picture the electron has to perform work against the field of the excited plasmon and thus loses a little bit of its energy. In [13] it is shown that a quantum mechanical description of EELS yields the same results as a semiclassical formalism if all the inelastic signal is collected [11]. This allows us to simply incorporate the relativistic motion of the electrons to our BEM approach[6] in terms of the *Liènard-Wiechert potentials*. We calculate the induced electric field $E_{ind}[r_e(t), \omega]$ at the positions of the electron beam r_e and can express the energy-loss probability through the work against this field according to [12, 15] as

$$\Gamma(\omega) = \frac{e}{\pi \hbar \omega} \int \Re e \left\{ e^{-i\omega t} \boldsymbol{v} \cdot \boldsymbol{E}_{ind}[r_e(t), \omega] \right\} \, dt . \qquad (5.5)$$

In the time domain the electron beam interacts with a surface plasmon oscillating in time, whereas in the frequency domain the plasmon oscillation becomes frozen and the electrons interact with a periodically modulated charge distribution as indicated in the expressions for the electron charge in Fig. 5.7. It is important to notice that we obtain a *nonlocal* interaction here, the electron at position r_e induces the plasmonic field which acts back on the electron at a later time at position r_e'. In this sense the energy loss probability can be expressed through the integral over the imaginary part of the induced Green tensor [16, 17] at the beam position $r_0 = (x_0, y_0)$

Electron energy loss probability in terms of Green function

$$\Gamma(r_0, \omega) \propto \int dt \, dt' \, \Im m \left\{ e^{i\omega(t'-t)} G^{ind}[r_e(t), r_e(t'); \omega] \right\}, \qquad (5.6)$$

which directly follows from Eq. (5.5). As discussed in [7] we calculated the energy loss for a rod-shaped particle in comparison to STEM EELS and EFTEM measurements performed by the FELMI group at the Graz University of Technology. In Fig. 5.8 we show the results of experiment and theory, which once again show intriguing correlations.

---

[6]An implementation of an electron-driven discrete-dipole approximation (e-DDA) can be found in [14].

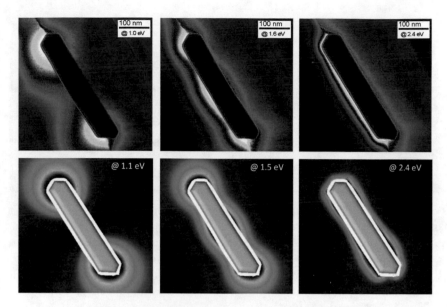

Fig. 5.8 Comparison of measured (*top*) and simulated (*bottom*) EFTEM maps as described in [7]. On the *left* we see the plasmon dipole mode with an energy of about 1 eV, in the *middle* the quadrupole mode at around 1.5 eV and the octupole mode at 2.4 eV on the *right*

Additional remark

Why is the kinetic energy of the electrons in EELS experiments so high?

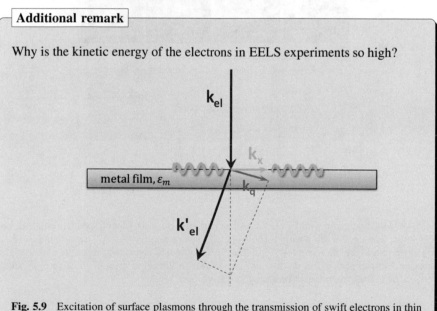

Fig. 5.9 Excitation of surface plasmons through the transmission of swift electrons in thin films [18, 19]. The incoming wave vector k_{el} transfers the component k_q to the thin film. The energy of the surface plasmon is determined by the projection k_x to the surface and the dispersion relation (2.10)

┌─ **Additional remark** ───┐

In fact, surface plasmons can also be excited by electron beams with lower
energies (for example reflection EELS experiments with energies of 1000–
2000 eV[7]). But in transmission EELS the electron has to pass through the
material and the smaller the electron wavelength the better the resolution, see
Fig. 5.9.

Note that for particle plasmons there is no momentum conservation, see
Chap. 2.

└───┘

Most of the time it is quite useful in understanding plasmonic spectra or
nanoscale processes if the eigenmodes of the corresponding nanoparticle are known.
As we have seen in Sect. 3.4.2, we can perform an eigenmode expansion to
mathematically obtain these modes. When it comes to mode mapping from the
experimental side, EELS turned out to be a very beneficial method since the swift
electrons also couple to optically dark modes and thus the full range of plasmon
modes can be depicted, see Fig. 5.10.

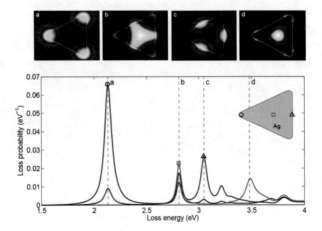

Fig. 5.10 Mode mapping with EELS for a silver nanotriangle, with friendly permission of Ulrich
Hohenester. The three beam positions (*blue circle*, *green rectangle*, and *red triangle*) indicated at
the inset generate the three spectral lines. The eigenmodes and eigenenergies of the nanotriangle
correspond to the peak positions **a–d** and are shown in the upper part of the figure. Panel **d**
corresponds to the optically dark breathing mode, see Sect. 3.4.2.1

---

[7]In 1956 Powell [20] investigated inconsistencies in the EELS spectra of transmission and
reflection EELS experiments which lead to the experimental verification of surface plasmons (also
see Fig. 2.1).

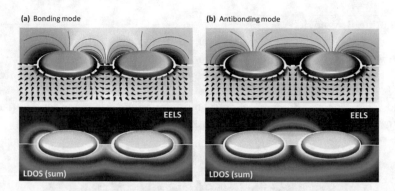

Fig. 5.11 The *lower panels* show EELS and LDOS maps for the bonding (**a**) and antibonding modes (**b**) of coupled nanoparticles. In the *upper panels* the corresponding scalar potential and electric field is plotted. As discussed in [16], the EELS signal is connected to the scalar potential whereas the photonic LDOS measures the electric field

There has also been some discussion about whether the EELS signal directly renders the photonic LDOS [16, 17]. It turns out that–although intimately related–there is no direct link between EELS and LDOS maps, and that EELS can even be blind to hot spots in the gap between coupled nanoparticles [16].

For the bonding plasmon mode plotted on the left side of Fig. 5.11 the fields in the gap region are very strong, resulting in a very high LDOS. The EELS signal on the contrary is completely blind to the hot spot. For the antibonding mode on the right hand side the behavior changes to the opposite: Because of the vanishing fields the LDOS drops in the gap, whereas the EELS rate has a pronounced maximum [16].

From Eq. (5.5) we also directly see that EELS is efficient only if the induced electric field is oriented along the electron beam path. The optical property of a dipole lying in a plane perpendicular to the electron beam can nevertheless be measured of course, since the induced field lines also point outwards of the dipole plane [12]. But we should keep in mind that certain modes, having part of their induced field lines perpendicular to the electron beam trajectory, may not be easy to map directly.

> ## Additional remark
>
> In this section we have discussed the interaction of high energetic electrons with metallic nanoparticles, where the impinging charges excite the plasmonic fields. However, the high and concentrated plasmonic near-fields can also be used for a reversed process, where photo-emitted electrons are accelerated away from the nanoparticle surface (further details and discussion see [21]).
>
>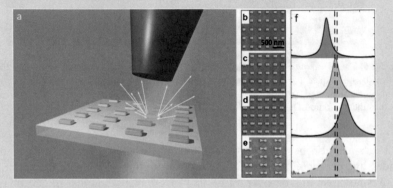
>
> **Fig. 5.12** (**a**) Scheme of hot electron emission from metallic nanoparticles. A femtosecond laser pulse excites a regular array of nanoantennas from below. In the strong plasmonic fields, electrons become photoemitted and ponderomotively accelerated and are finally analyzed by time-of-flight spectrometry. (**b**)–(**e**) SEM images of nanoparticle arrays which are blue-shifted, in resonance, and red-shifted with respect to the excitation bandwidth (see *dashed lines* in the spectra). (**f**) Measured extinction spectra for the corresponding particles. Figure adapted with permission from [21], © 2013 American Chemical Society
>
> The electrons become photoexcited either through multi-photon absorption or quantum-mechanical tunneling and since the electric field varies strongly along the surface of the nanoparticle, both mechanisms need to be taken into account (see Fig. 5.12). The simulation of the electron trajectories can then be accomplished within the so-called *simple-man model*, where the emitted electrons become ponderomotively accelerated in the total field of the nanoparticle. This photoacceleration process is governed by the evanescent surface plasmon field of the nanoparticle, which allows for a high-level control of electron emission by tailoring the geometry and thereby the plasmonic particle resonances. This offers unique prospects for the generation and all-optical control of plasmonic electron sources as well as other applications in lightwave electronics.

5.2.3 Plasmon Tomography

A detailed comparison between the photonic LDOS and EELS was given recently in [22], where the authors provided an intuitive interpretation of different measurement schemes in terms of an eigenmode expansion. We have recently shown [23] that for sufficiently small nanoparticles, where the quasistatic approximation can be employed, an expansion of the particle fields in terms of these plasmonic eigenmodes also allows a different approach for the interpretation of EELS data: The EELS signal can be considered as a spatial average of the eigenmode potential along the electron propagation direction, and the extraction of plasmon fields from EELS data can be reduced to an inverse *Radon transformation*,[8] which is at the heart of most modern computer tomography algorithms [25–30]. The Shepp-Logan phantom [31] plotted in Fig. 5.13 serves as a standard model image of a human head in testing various image reconstruction algorithms. The Radon transform of the original image Fig. 5.13a is called *sinogram*[9] and shown in panel (b). In (c) the reconstructed image after the inverse transform is drawn.

The basic principle of the Radon transform $\mathcal{R}$ is explained in Fig. 5.14 again for the Shepp-Logan phantom. The intensity profile for a certain angle θ_1 in Fig. 5.14a is generated by integrating along each projection line (dashed arrows). If we hold certain positions indicated as red, blue, purple, and orange dots in (a) and vary the angle θ we obtain the corresponding colored lines in panel (b). A complete projection set for $-90° < \theta < 90°$ is plotted in panel (c) and represents the Radon transform of the Shepp-Logan phantom. It is extraordinary that we can apply the same reconstruction to EELS measurements and extract abstract mathematical quantities from experimental data. The projection line is then represented by the

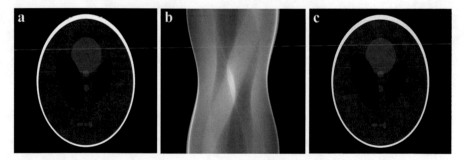

Fig. 5.13 Principle of the Radon reconstruction for the Shepp-Logan phantom test image. Panel (**a**) shows the original image, (**b**) the sinogram of the corresponding Radon transform, and panel (**c**) the reconstruction of (**b**) after applying the inverse transform

---

[8]The method was introduced by the Austrian mathematician Johann Radon in 1917 [24].

[9]The name comes from the fact that the Radon transform of a Dirac delta function is linked to sinusoidal curves.

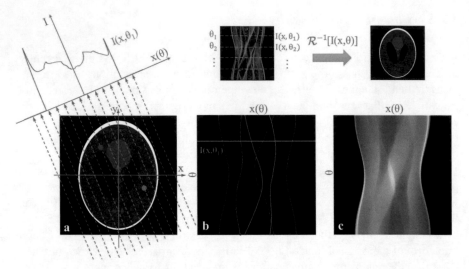

Fig. 5.14 Principle of the Radon transformation as described in the text. (**a**) Intensity profiles are produced by integrating along each projection line for a certain tilting angle θ. Holding several positions indicated as colored dots in (**a**) and varying θ generates the corresponding lines in (**b**). The *green horizontal line* corresponds to the intensity image for θ_1 in (**a**). The complete Radon transform for $-90° < \theta < 90°$ in steps of $2°$ is plotted in the sinogram (**c**). The inset above shows the reconstruction by applying the inverse Radon transform $\mathcal{R}^{-1}$

electron beam and the intensity profile by the averaged eigenmode potential. In Eq. (5.6) we have seen that the loss probability for a certain impact parameter r_0 can be expressed in terms of the quasistatic Green function G^{ind} as

$$\Gamma(r_0, \omega) \propto \int dt\, dt'\, \Im\left\{ e^{i\omega(t'-t)} G^{\text{ind}}[r_e(t), r_e(t'); \omega] \right\}.$$

The Green function is a quantity of central interest in the BEM approach and by following [23] the decomposition in quasistatic eigenmodes with corresponding eigenvalues λ_k leads to [22]

$$G^{\text{ind}}(r, r') = -\sum_k \frac{\lambda_k \pm 2\pi}{\Lambda + \lambda_k} \phi_k(r)\phi_k(r') \frac{1}{\varepsilon(r')}, \qquad (5.7)$$

where $\phi_k = \int da\, G\sigma_k$ (assumed to be real-valued) is the potential of the kth eigenmode. The dielectric functions inside and outside the particle are ε_1 and ε_2, respectively, and also the matrix $\Lambda = 2\pi \frac{\varepsilon_2 + \varepsilon_1}{\varepsilon_2 - \varepsilon_1} \mathbb{1}$ from Sect. 3.4.1 has been used once more. The plus and minus signs in Eq. (5.7) correspond to the situations where r' lies outside or inside the particle. If we insert Eq. (5.7) into the expression of Γ,

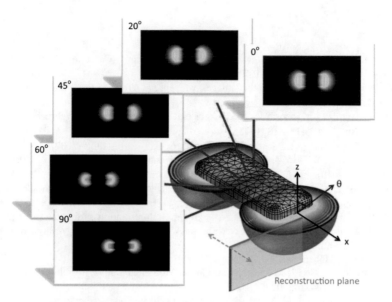

Fig. 5.15 Schematics of EELS tomography, reprinted with permission from [23], © 2013 by The American Physical Society. An electron beam is raster scanned over a metallic nanoparticle, and EELS maps are recorded for different rotation angles. The main panel shows the isosurface and contour lines for the modulus of the dipolar surface plasmon potential, and the insets report the different EELS maps. From the complete collection of maps one can reconstruct the plasmon fields

we end up with the quasistatic eigenmode expansion of the loss probability. At a plasmon resonance the EELS probability then reduces to [23]

$$\Gamma_\theta \propto |\mathcal{R}_\theta[\phi_k(\boldsymbol{r})]|^2\,, \tag{5.8}$$

where $\mathcal{R}_\theta$ is the Radon transform for a certain rotation angle θ that performs a line integration of ϕ_k along the z direction. The *projection-slice theorem* states that one can uniquely reconstruct the original function from a collection of Radon transformations for a complete set of rotation angles, see Fig. 5.15.

This reconstruction method for the quasistatic regime has been demonstrated in [23] and [30]. To expand the formalism to larger nanostructures too, where retardation effects become important and the quasistatic approximation fails, a clear concept for retarded eigenmodes is essential. The development of such a framework is part of our current work, see for example [36].

References

1. L. Novotny, B. Hecht, *Principles of Nano-Optics*, 2nd edn. (Cambridge University Press, Cambridge, 2012). ISBN 978-1107005464
2. L. Rayleigh, On the theory of optical images, with special reference to the microscope. Philos. Mag. **42**, 167–195 (1896).
3. E. Abbe, Beiträge zur Theorie des Mikroskops und der mikroskopischen Wahrnehmung. Arch. Mikrosk. Anat. **9**(1), 413–418 (1873).
4. S.W. Hell, Far-field optical nanoscopy. Science **316**(5828), 1153–1158 (2007).
5. D.K. Gramotnev, S.I. Bozhevolnyi, Plasmonics beyond the diffraction limit. Nat. Photonics **4**(2), 83–91 (2010).
6. U. Hohenester, A. Trügler, Interaction of single molecules with metallic nanoparticles. IEEE J. Sel. Top. Quantum Electron. **14**, 1430 (2008).
7. B. Schaffer, U. Hohenester, A. Trügler, F. Hofer, High-resolution surface plasmon imaging of gold nanoparticles by energy-filtered transmission electron microscopy. Phys. Rev. B **79**, 041401(R) (2009).
8. D. Koller, U. Hohenester, A. Hohenau, H. Ditlbacher, F. Reil, N. Galler, F. Aussenegg, A. Leitner, A. Trügler, J. Krenn, Superresolution Moire mapping of particle plasmon modes. Phys. Rev. Lett. **104**, 143901 (2010).
9. M. Bosman, V.J. Keast, M. Watanabe, A.I. Maaroof, M.B. Cortie, Mapping surface plasmons at the nanometre scale with an electron beam. Nanotechnology **18**, 165505 (2007).
10. J. Nelayah, M. Kociak, O. Stephan, F.J. García de Abajo, M. Tence, L. Henrard, D. Taverna, I. Pastoriza-Santos, L.M. Liz-Martin, C. Colliex, Mapping surface plasmons on a single metallic nanoparticle. Nat. Phys. **3**, 348 (2007).
11. F.J. García de Abajo, Optical excitations in electron microscopy. Rev. Mod. Phys. **82**, 209 (2010).
12. M. Kociak, O. Stéphan, Mapping plasmons at the nanometer scale in an electron microscope. Chem. Soc. Rev. **43**, 3865–3883 (2014).
13. R.H. Ritchie, A. Howie, Inelastic scattering probabilities in scanning transmission electron microscopy. Philos. Mag. A **58**(5), 753–767 (1988).
14. N.W. Bigelow, A. Vaschillo, V. Iberi, J.P. Camden, D.J. Masiello, Characterization of the electron- and photon-driven plasmonic excitations of metal nanorods. ACS Nano **6**(8), 7497–7504 (2012).
15. F.J. García de Abajo, A. Howie, Retarded field calculation of electron energy loss in inhomogeneous dielectrics. Phys. Rev. B **65**, 115418 (2002).
16. U. Hohenester, H. Ditlbacher, J.R. Krenn, Electron-energy-loss spectra of plasmonic nanoparticles. Phys. Rev. Lett. **103**, 106801 (2009).
17. F.J. García de Abajo, M. Kociak, Probing the photonic local density of states with electron energy loss spectroscopy. Phys. Rev. Lett. **100**, 106804 (2008).
18. H. Raether, *Surface Plasmons on Smooth and Rough Surfaces and on Gratings*. Springer Tracts in Modern Physics, vol. 111 (Springer, Berlin, 1988). ISBN 978-0387173634
19. H. Raether, *Excitation of Plasmons and Interband Transitions by Electrons*. Springer Tracts in Modern Physics, vol. 88 (Springer, Berlin, 1980). ISBN 978-3540096771
20. C.J. Powell, J.B. Swan, Origin of the characteristic electron energy losses in aluminum. Phys. Rev. **115**, 869 (1959).
21. P. Dombi, A. Hörl, P. Rácz, I. Márton, A. Trügler, J.R. Krenn, U. Hohenester, Ultrafast strong-field photoemission from plasmonic nanoparticles. Nano Lett. **13**(2), 674–678 (2013).
22. G. Boudarham, M. Kociak, Modal decompositions of the local electromagnetic density of states and spatially resolved electron energy loss probability in terms of geometric modes. Phys. Rev. B **85**, 245447 (2012).

23. A. Hörl, A. Trügler, U. Hohenester, Tomography of particle plasmon fields from electron energy loss spectroscopy. Phys. Rev. Lett. **111**, 076801 (2013).
24. J. Radon, Über die Bestimmung von Funktionen durch ihre Integralwerte längs gewisser Mannigfaltigkeiten. Akad. Wiss. **69**, 262–277 (1917)
25. G.T. Herman, *Image Reconstruction from Projections: The Fundamentals of Computerized Tomography* (Academic, New York/London, 1980). ISBN 978-1-84628-723-7
26. A.C. Twitchett-Harrison, T.J.V. Yatesa, R.E. Dunin-Borkowski, P.A. Midgleya, Quantitative electron holographic tomography for the 3D characterisation of semiconductor device structures. Ultramicroscopy **108**, 1401–1407 (2008).
27. N. Jin-Phillipp, C. Koch, P. van Aken, Toward quantitative core-loss EFTEM tomography. Ultramicroscopy **111**, 1255–1261 (2011).
28. R. Leary, Z. Saghi, P.A. Midgley, D.J. Holland, Compressed sensing electron tomography. Ultramicroscopy **131**, 70–91 (2013).
29. J.M. Thomas, R. Leary, P.A. Midgley, D.J. Holland, A new approach to the investigation of nanoparticles: electron tomography with compressed sensing. J. Colloid Interface Sci. **392**, 7–14 (2013).
30. O. Nicoletti, F. de la Pena, R.K. Leary, D.J. Holland, C. Ducati, P.A. Midgley, Three-dimensional imaging of localized surface plasmon resonances of metal nanoparticles. Nature **502**, 80–84 (2013).
31. L.A. Shepp, B.F. Logan, The fourier reconstruction of a head section. IEEETrans. Nucl. Sci. **21**(3), 21–43 (1974).
32. X. Zhou, A. Hörl, A. Trügler, U. Hohenester, T.B. Norris, A.A. Herzing, Effect of multipole excitations in electron energy-loss spectroscopy of surface plasmon modes in silver nanowires. J. Appl. Phys. **116**, 233101 (2014).
33. F.P. Schmidt, H. Ditlbacher, A. Trügler, U. Hohenester, A. Hohenau, F. Hofer, J.R. Krenn, Plasmon modes of a silver thin film taper probed with STEM-EELS. Opt. Lett. **40**, 5670–5673 (2015).
34. G. Haberfehlner, A. Trügler, F.P. Schmidt, A. Hörl, F. Hofer, U. Hohenester, G. Kothleitner, Correlated 3D nanoscale mapping and simulation of coupled plasmonic nanoparticles. Nano Lett. **15**, 7726–7730 (2015).
35. C. Cherqui, N. Thakkar, G. Li, J.P. Camden, D.J. Masiello, Characterizing localized surface plasmons using electron energy-loss spectroscopy. Annu. Rev. Phys. Chem. **67**(1) (2016).
36. A. Hörl, A. Trügler, U. Hohenester, Full three-dimensional reconstruction of the dyadic Green tensor from electron energy loss spectroscopy of plasmonic nanoparticles. ACS Photonics **2**, 1429 (2015).

Chapter 6
Influence of Surface Roughness

> *God made the bulk; surfaces were invented by the devil.*
>
> WOLFGANG PAULI

It is nearly impossible to fabricate perfectly smooth nanoparticles, therefore it is important to discuss the influence of surface roughness on the optical properties of MNPs [1, 34]. Especially if e-beam lithography (see Sect. 2.10.2) is used to produce the particles, the resulting metal structures are polycrystalline and the surfaces are quite rough [2, 3]. Contrary to what one might anticipate, initial indications are that a moderate amount of surface roughness has no significant impact on the optical properties of MNP, at least in the far-field region. In [4] we show that this behavior can be interpreted as some kind of plasmonic averaging over the random height fluctuations of the rough metal surface (*motional narrowing*).

If the properties of MNPs are investigated in the near-field region, however, the influence of roughness features or lift-off artifacts can become important. Local variations or exceptional hot-spots of the electromagnetic near-field may lead to detectable deviations compared to smooth particles, especially when surface sensitive methods like SERS come into play. Usually the surface roughness of a nanoparticle can be reduced by thermal annealing,[1] but this also influences the dielectric response of the metal. In general after such a heating process a complex interplay of the reduced nanoscopic roughness, average crystallite grain sizes, dielectric and morphological changes can be observed, see [5] and Sect. 6.3.

6.1 Generation of a Rough Particle in the Simulation

To allow systematic investigations of surface roughness on metallic nanoparticles, we need to control and quantify the "amount" of roughness on a surface. One possibility to do this is to add (controllable) stochastic height variations to the smooth surface of an ideal nanoparticle. Let us exemplify the idea for the 2D case: We assume that the height of each vertex of a triangulated plate is given by a function

---

[1]Typically the particle is tempered for a couple of minutes at temperatures around 200°C on a hot plate.

© Springer International Publishing Switzerland 2016
A. Trügler, *Optical Properties of Metallic Nanoparticles*, Springer Series in Materials Science 232, DOI 10.1007/978-3-319-25074-8_6

$h(x, y)$. The mean value should vanish $\langle h(x, y) \rangle = 0$ and the value of h at (x', y') should be correlated to the value at position (x, y) through a distance dependent function: $\langle h(x, y)h(x', y') \rangle = f(x - x', y - y')$. These requirements are all fulfilled if we introduce a Gaussian autocorrelation function and attach arbitrary phase factors to all Fourier coefficients. In this way we are able to average the z-coordinate of each vertex corresponding to this random potential and we have two parameters to control the roughness, the standard deviation of the Gaussian function ς (determines the amount of roughness) and a height scaling parameter Δh (assigns the deepness of the asperity).

Wrapping the resulting roughness around the surface of a nanorod, we obtain the structure shown in Fig. 6.1.

Unfortunately very sharp features in the triangulated mesh may lead to inaccuracies in the simulation and may cause diverging surface charge densities. Luckily we obtain an even better result without sharp peaks, if we do not directly interpolate the height of each vertex with the Gaussian autocorrelation, but rather generate a box around the particle and interpolate the stochastic vertex height with respect to this box, see Fig. 6.2.

Again we use a normal probability density function multiplied with N random phases $e^{i\phi_{rnd}}$ (for each spatial direction)

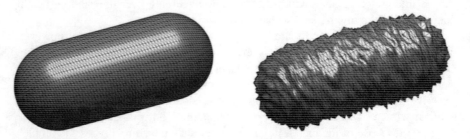

Fig. 6.1 Triangulated nanorod with smooth surface and stochastic height variations generated with a random Gaussian potential. Problems in the simulation may arise if sharp features are present

Fig. 6.2 Generating a box around the smooth particle and interpolating stochastic vertex heights with respect to this box gives a much better result without sharp features as in Fig. 6.1. The random height variations are again controllable by parameters of the Gaussian autocorrelation, see Eq. (6.1)

Generation of surface roughness

$$h(x, y, z) = \frac{1}{N} \, \Re e \left\{ \mathcal{F}^{-1} \left[e^{-\frac{1}{2}\varsigma^2 \left(k_x^2 + k_y^2 + k_z^2\right)} \, e^{i\phi_{\mathrm{rnd}}} \right] - \frac{1}{2} \right\}, \tag{6.1}$$

where $\mathcal{F}^{-1}$ denotes the inverse Fourier transform, N is a normalization factor and ς^2 is the variance of the height fluctuations. For the arbitrary phase factors ϕ_{rnd} the MATLAB® built-in function rand(n) to create uniformly distributed pseudorandom numbers has been used.

We next interpolate the scaled stochastic height variations $\Delta h \cdot h(x, y, z)$ to the nanoparticle surface and displace the vertices of the nanoparticle along the surface normal directions. The influence of Δh and ς is shown in Fig. 6.3.

Another possibility for the simulation of "realistic" particles is to be found in the extraction of their contour out of scanning electron microscope (SEM) images. The triangulated particle is generated by assigning a certain height profile (with a rounded transition to the top and bottom area) to the extracted contour-polyline and wrapping a mesh around the structure. A SEM image together with the extracted contour and the final BEM particle is shown in Fig. 6.4.

Constant height scaling Δh, varying variance ς

Constant variance ς, varying height scaling Δh

Fig. 6.3 Surface roughness generated with Eq. (6.1) for changing parameters

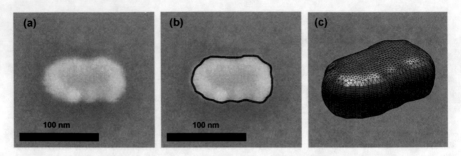

Fig. 6.4 (a) SEM image of a rod-shaped nanoparticle fabricated by e-beam lithography. (b) Contour extraction and (c) creation of the corresponding triangulated particle for a BEM simulation

If certain lift-off features or other local surface irregularities should be modeled as well, it is rarely possible to avoid a manual adjustment of the involved vertex positions.

6.2 Theoretical Analysis of Surface Roughness

Let us for one moment take only the mean value of the height difference of bumps or hollows at a rough surface as an indicator for the degree of roughness. For a medium sized nanoparticle this mean value is of the order of nanometers. One possibility to describe the optical properties of structures with such small features[2] is presented in [7], for example. By writing the dielectric constant $\varepsilon(\mathbf{k},\omega)$ as a function of both the wave vector $\mathbf{k}$ and the angular frequency ω, the authors incorporate the effect of small features via the spatially nonlocal response of materials to Maxwell's equations[3] (see Chap. 8).

On the other hand we can also understand the influence of rough surfaces if we once again consider the eigenmode expansion discussed in Sect. 3.4.2. We can model the rough surface as a distortion of the surface $\partial\Omega$ from its ideal shape, see [4]. The surface derivative of the Green function F then changes to $F+\delta F$.

We want to determine, how the peak position of the plasmons is affected in case of surface roughness and since this peak position is assigned by the plasmon energy, we need to investigate the modification of the Eigenvalue λ_k. For sufficiently small δF we can employ a perturbation analysis, where we treat F as the unperturbed part and δF as the "perturbation". This calculation is carried out in [4].

In the end we obtain a surprisingly small effect for the influence of surface roughness on the position and width of the plasmon peaks of metallic nanoparticles. The reason for this is motional narrowing,[4] where the plasmon averages over the random height fluctuations h.

---

[2]So-called finite size effects have been phenomenologically accounted for by increasing the damping rate of the conduction electrons contribution to the permittivity, see [6].

[3]Their implementation is based on the self-consistently solved hydrodynamic Drude model.

[4]This behavior is known from electron-hole pairs in semiconductor quantum wells [8], where the propagating excitons "average" over the random potential of local monolayer fluctuations, which results in a narrowing of the exciton lineshape.

6.3   Near-Field Consequences of Rough Nanoparticles

The situation changes if we are interested in the near-field regime, especially for SERS, where the quality of the nanoparticle surface is known to play a crucial role. The SERS enhancement is proportional to the fourth power of the field intensity [9], see Eq. (2.14), thus lithographic lift-off artifacts or other nanoscopic roughness features can lead to strong local variations.

In [5] we show that a major contribution to the optical near-field and SERS enhancement results from typical edge-roughness features (local protrusion of 20 nm extension), provided that they are situated at regions of large plasmonic mode density. Although they only occupy a small surface region of the nanoparticle, the contribution can dominate even the average SERS signal. In contrast, surface roughness on the top side of a nanoparticle gives only small contributions to the SERS enhancement and has almost no impact on the averaged SERS signal. As mentioned at the beginning of this chapter, a possibility to reduce the surface roughness of a lithographic particle is thermal annealing, see Fig. 6.5. Besides the morphological changes of the lithographic, polycrystalline particle this also modifies the average crystallite grain size and the corresponding dielectric function. On one hand the reduction of surface roughness reduces the accompanying SERS enhancement but on the other it also enhances the SERS signals due to lower Ohmic damping of the plasmon resonance. Depending on the relative importance of these various effects on an actual sample under investigation, annealing can either lead to a reduction of the average SERS signal (as usually observed experimentally) or

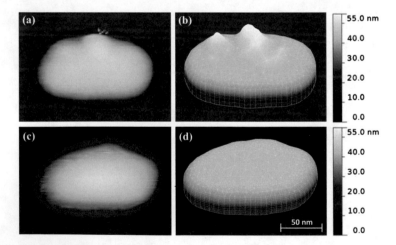

Fig. 6.5 (a) AFM image of a lithographic nanoparticle. (b) Reconstructed particle geometry with surface mesh for simulation. (c) Same particle as in (a), but after a 200°C annealing process. The thermal curing yields reduced surface roughness but also affects the crystallite grain size and dielectric function (details see [5]). This causes a spectral shift with respect to the original particle and modifies the near field enhancement. (d) Reconstruction of the annealed particle

lead to an enhancement (to be expected in the red spectral range, on lithographic arrays with little or no edge-roughness features). Hence in the end, the SERS signal of an array of lithographic particles critically depends on the exact location and total number of roughness features, as well as the particles' dielectric function.

References

1. H. Raether, *Surface Plasmons on Smooth and Rough Surfaces and on Gratings*. Springer Tracts in Modern Physics, vol. 111 (Springer, Berlin, 1988). ISBN 978-0387173634
2. J. Rodríguez-Fernández, A.M. Funston, J. Pérez-Juste, R.A. Álvarez-Puebla, L.M. Liz-Marzán, P. Mulvaney, The effect of surface roughness on the plasmonic response of individual submicron gold spheres. Phys. Chem. Chem. Phys. **11**, 5909 (2009).
3. K.-P. Chen, V.P. Drachev, J.D. Borneman, A.V. Kildishev, V.M. Shalaev, Drude relaxation rate in grained gold nanoantennas. Nano Lett. **10**, 916 (2009).
4. A. Trügler, J.-C. Tinguely, J.R. Krenn, A. Hohenau, U. Hohenester, Influence of surface roughness on the optical properties of plasmonic nanoparticles. Phys. Rev. B **83**, 081412(R) (2011).
5. A. Trügler, J.-C. Tinguely, G. Jakopic, U. Hohenester, J.R. Krenn, A. Hohenau, Near-field and SERS enhancement from rough plasmonic nanoparticles. Phys. Rev. B **89**, 165409 (2014).
6. U. Kreibig, M. Vollmer, *Optical Properties of Metal Clusters. Springer Series in Material Science*, vol. 25 (Springer, Berlin, 1995). ISBN 978-3-540-57836-9
7. J.M. McMahon, S.K. Gray, G.C. Schatz, Nonlocal optical response of metal nanostructures with arbitrary shape. Phys. Rev. Lett. **103**, 097403 (2009).
8. M.Z. Maialle, E.A. de Andrada e Silva, L.J. Sham, Exciton spin dynamics in quantum wells. Phys. Rev. B **47**, 15 776 (1993).
9. E.C. Le Ru, J. Grand, N. Félidj, J. Aubard, G. Lévi, A. Hohenau, J.R. Krenn, E. Blackie, P.G. Etchegoin, Experimental verification of the SERS electromagnetic model beyond the $|E|^4$ approximation: polarization effects. J. Phys. Chem. C **112**, 8117–8121 (2008).
10. G. Haberfehlner, A. Trügler, F.P. Schmidt, A. Hörl, F. Hofer, U. Hohenester, G. Kothleitner, Correlated 3D nanoscale mapping and simulation of coupled plasmonic nanoparticles. Nano Lett. **15**, 7726–7730 (2015).

Chapter 7
Nonlinear Optical Effects of Plasmonic Nanoparticles

> *But the speed was power, and the speed was joy, and the speed was pure beauty*
>
> RICHARD BACH, Jonathan Livingston Seagull

The light-matter interaction between metallic nanoparticles and an electromagnetic wave happens on a very fast time scale and, as we have discussed in Sect. 2.7, already after a few femtoseconds the plasmonic excitations start to vanish again. The temporal evolution and this ultrafast relaxation of surface plasmon polaritons is of central importance for many kinds of plasmonic applications. It is amazing that we can explore such rapid dynamic processes with experiments nowadays. One way to enter the ultrafast world of plasmon dephasing is given by nonlinear autocorrelation measurements, which allow us to determine sub-10 fs decay times. Usually a bandwidth-limited laser pulse working in the few-cycle regime is used to excite nonlinear effects which serve as a non-invasive monitor for the plasmon dynamics [1].

The simplest nonlinear effect is second-harmonic generation (SHG), where a nonlinear material absorbs two photons of frequency ω and emits a new photon of frequency 2ω. But because of the inversion symmetry in the atomic arrangement of metals, SHG is forbidden in the bulk [1–3]. However the symmetry is broken at the surface and so a second order dipole response can exist in the surface region. There can be electric dipolar surface contributions due to the broken symmetry or also higher order bulk signals can contribute to the second-order nonlinearity [1], but the latter are known to be much weaker and can therefore be neglected.

Breaking the symmetry of centrosymmetric media also allows a second-order signal for a perfectly symmetric nanosphere. For example, it will strongly depend on defects, facets and other small deviations from the spherical shape [4]. This defect dependence will give a huge varying signal for different particles of the same shape. Another example of SHG measurements for systems with a broken symmetry has been reported by [5], where they investigated arrays of noncentrosymmetric T-shaped gold nanodimers.

On the other hand third-harmonic generation (THG), the next higher nonlinear process where a mixing of four fields $\left(E \propto \chi^{(3)} E^3\right)$ occurs, is allowed in all media and not restricted to symmetry considerations, see Fig. 7.1. Supported by the above mentioned reasons the $\chi^{(3)}$ nonlinearity for gold nanoparticles is also much stronger

© Springer International Publishing Switzerland 2016

A. Trügler, *Optical Properties of Metallic Nanoparticles*, Springer Series in Materials Science 232, DOI 10.1007/978-3-319-25074-8_7

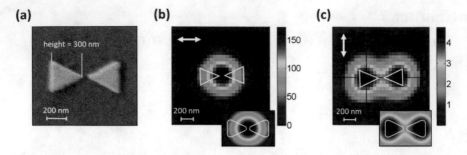

Fig. 7.1 (**a**) SEM image of a gold bowtie nanoantenna. (**b**) and (**c**) measured THG intensity (kcts/s) for linear polarization along long and short axes respectively. The lower *insets* show the corresponding simulation results, for further details see [6]. (Measurements by Tobias Hanke, University of Konstanz)

than the corresponding SHG signal [7], which makes THG a perfect candidate to investigate the femtosecond dynamics of particle plasmons.

The first observation of third-harmonic signals from individual gold colloids down to 40 nm diameter can be found in [4] and a phenomenological macroscopic theory of optical second- and third-harmonic generation from cubic centrosymmetric crystals has been discussed by [8].

7.1   Autocorrelation

To measure the autocorrelation a sample is excited by two sequenced laser pulses with a varying time delay t' between the pulses. The collected intensity is given by the interfering electric fields to the power of four (second order autocorrelation G_2) or to the power of six (third order autocorrelation G_3) and averaged over the detector response time T [1]:

Second and third order autocorrelation function

$$\text{SHG:} \qquad G_2(t') = \frac{1}{T} \int_{-T}^{+T} \left[E(t) + E(t - t') \right]^4 \, dt, \qquad (7.1)$$

$$\text{THG:} \qquad G_3(t') = \frac{1}{T} \int_{-T}^{+T} \left[E(t) + E(t - t') \right]^6 \, dt. \qquad (7.2)$$

This pulse interference can also be described by a simple harmonic oscillator model (see [1]). By inserting harmonic fields in the above expressions for G_2 or G_3 the characteristic peak ratio of 1:8 for SHG or 1:32 for THG follows directly from the in and out of phase expression of the electric fields. In Fig. 7.2 the measured and simulated result of G_3 for three different nanoantennas is shown.

Fig. 7.2 Third order autocorrelation for three different nanoantenna geometries, see [6] for spatial dimensions and further details. (**a**) Experimental and (**b**) simulation result. (Figure with friendly permission by Tobias Hanke, University of Konstanz)

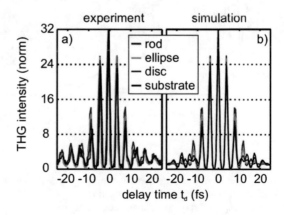

7.2   Third Harmonic Imaging

In [6] we discuss the third harmonic emission of five different antenna structures. The THG intensity differs significantly for the different structures and we show that the knowledge of the plasmon damping time (obtained from autocorrelation measurements) suffices to explain the differences. The principal scheme for this experiment is depicted in Fig. 7.3. An array of nanoantennas (varying in size with different gap distances) is illuminated by a laser pulse and the nonlinear emission intensity is collected.

The dielectric environment of the metallic particles can also contribute to the nonlinear emission, but the susceptibility of air or fused silica is at least several orders of magnitude weaker than that of gold (see Sect. 1.3 for the unit conversion):

$$\chi_e^{(3)} \approx 10^{-11} \text{ [esu]} = 1.4 \times 10^{-19} \left[\text{m}^2/\text{v}^2\right] \text{ for gold [9],}$$

$$\chi_e^{(3)} \approx 10^{-14} \text{ [esu]} = 1.4 \times 10^{-22} \left[\text{m}^2/\text{v}^2\right] \text{ for fused silica [10],}$$

$$\chi_e^{(3)} \approx 10^{-18} \text{ [esu]} = 1.4 \times 10^{-26} \left[\text{m}^2/\text{v}^2\right] \text{ for air [11].}$$

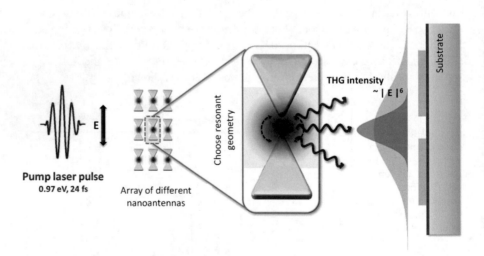

Fig. 7.3 Principle of third harmonic imaging as described in [6]. The few-fs pulse of a pump laser excites a nanoantenna and the nonlinear emission gets collected. By raster scanning the laser pulse over a sample of varying antenna geometries the most resonant structure for this excitation can be picked, see [6]

Fig. 7.4 Schematic representation of a bowtie nanoantenna used as one of five different antenna geometries in [6]. One half of the antenna shows the surface charge distribution at the resonance energy (the *bonding mode* is clearly visible), whereas the other half depicts the surface discretization needed for the simulation with the BEM

Nonlinear emission spectroscopy emerges as a new powerful tool for the spatiotemporal characterization of nanoantennas. As described in [6], it is not the shape but the volume of the nanoantennas that plays a crucial role for the determination of the nonlinear emission intensity. Due to radiative damping the structure with the lowest active volume generates by far the strongest THG emission.

In [4] it has been stated that THG signals also serve as label for bio-sensors and may be used for the tracking of single molecules. Placing an emitter in the gap region of the nanoantennas (see Fig. 7.4) may also be one of the next steps to reach this goal (also see [12]).

A recently published review about nonlinear plasmonics can be found in [13].

References

1. B. Lamprecht, Ultrafast plasmon dynamics in metal nanoparticles. Ph.D. thesis, Institut für Physik, Karl-Franzens-Universität Graz, 2000.
2. P.S. Pershan, Nonlinear optical properties of solids: energy considerations. Phys. Rev. **130**, 919–929 (1963).
3. E. Adler, Nonlinear optical frequency polarization in a dielectric. Phys. Rev. **134**, A728–A733 (1964).
4. M. Lippitz, M.A. van Dijk, M. Orrit, Third-harmonic generation from single gold nanoparticles. Nano Lett. **5**, 799–802 (2005).
5. B.K. Canfield, H. Husu, J. Laukkanen, B. Bai, M. Kuittinen, J. Turunen, M. Kauranen, Local field asymmetry drives second-harmonic generation in noncentrosymmetric nanodimers. Nano Lett. **7**, 1251–1255 (2007).
6. T. Hanke, J. Cesar, V. Knittel, A. Trügler, U. Hohenester, A. Leitenstorfer, R. Bratschitsch, Tailoring spatiotemporal light confinement in single plasmonic nanoantennas. Nano Lett. **12**(2), 992–996 (2012).
7. T. Hanke, G. Krauss, D. Träutlein, B. Wild, R. Bratschitsch, A. Leitenstorfer, Efficient nonlinear light emission of single gold optical antennas driven by few-cycle near-infrared pulses. Phys. Rev. Lett. **103**, 257404 (2009).
8. J.E. Sipe, D.J. Moss, H.M. van Driel, Phenomenological theory of optical second- and third-harmonic generation from cubic centrosymmetric crystals. Phys. Rev. B **35**, 1129–1141 (1987).
9. N. Bloembergen, W.K. Burns, M. Matsuoka, Reflected third harmonic generated by picosecond laser pulses. Opt. Commun. **1**, 195–198 (1969).
10. U. Gubler, C. Bosshard, Optical third-harmonic generation of fused silica in gas atmosphere: absolute value of the third-order nonlinear optical susceptibility $\chi^{(3)}$. Phys. Rev. B **61**, 10702–10710 (2000).
11. F. Krausz, E. Wintner, Atmospheric influences in optical third-harmonic generation experiments. App. Phys. B **49**, 479–483 (1989).
12. A. Kinkhabwala, Z. Yu, S. Fan, Y. Avlasevich, K. Müllen, W.E. Moerner, Large single-molecule fluorescence enhancements produced by a bowtie nanoantenna. Nat. Photonics **3**, 654–657 (2009).
13. M. Kauranen, A.V. Zayats, Nonlinear plasmonics. Nat. Photonics **6**(11), 737–748 (2012).

Chapter 8
Nonlocal Response

> *I don't demand that a theory correspond to reality because I don't know what it is. Reality is not a quality you can test with litmus paper. All I'm concerned with is that the theory should predict the results of measurements*
>
> STEPHEN HAWKING

The response of metallic nanostructures to an external excitation has been introduced in this book through a semiclassical top-down approach, where the microscopic dynamic of the electrons has been lumped into the macroscopic dielectric function. We have discussed the classical concept of this function and in Sect. 3.3 we have seen that a simple frequency dependence of ε has far-reaching consequences like temporal nonlocal equations and the mandatory accompaniment of electromagnetic losses. In this formalism based on Maxwell's equations we also assume sharp boundary conditions and abrupt interfaces, which of course in the quantum world becomes a questionable approximation. We call this entire approach *spatially local*, since the dielectric function is assumed to be isotropic and only depends on the frequency ω. But as the control over size and morphology of fabricated nanostructures is pushed to the nanometer scale, the validity of this local description becomes more and more strained.

Although the quantum approach is more fundamental, the semiclassical description in terms of electromagnetic fields and a local dielectric response function works extremely well and has been successfully applied to a vast number of plasmonic applications and problems. Also the quantization of dissipative modes cannot be done in the usual straight forward textbook-like way, and the huge amount of electrons in a typical sized nanoparticle renders a quantum mechanical description often cumbersome, such that from the start additional approximations are required. In fact, the collective excitation of only 500–1000 conduction electrons is sufficient, to still form a classical plasmon resonance [1]. If the particles get even smaller than that, the broad spectral resonance peak splits into a band of discrete modes, with some having the collective character of the classical surface plasmon and others being more strongly localized in the core of the particle [1, 2]. In Fig. 8.1 such a

© Springer International Publishing Switzerland 2016
A. Trügler, *Optical Properties of Metallic Nanoparticles*, Springer Series in Materials Science 232, DOI 10.1007/978-3-319-25074-8_8

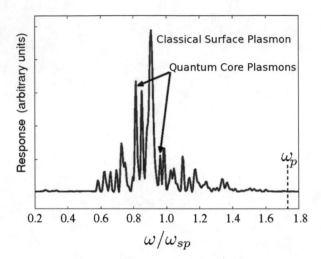

Fig. 8.1 Optical spectrum of a single 100 electron Au-jellium sphere with a radius of 0.74 nm. The x-axis indicates frequency in units of the classical surface plasmon frequency, ω_{sp}. The bulk plasmon frequency, ω_p, is also indicated. Reprinted with permission from [2]. © 2012 American Chemical Society

spectrum is plotted for a jellium[1] sphere with $r = 0.74$ nm (≈ 100 electrons). If we end up with just a few metal atoms and change from nanoparticles to a molecular cluster, we finally obtain transitions between discrete states, very similar to the transitions in isolated atoms or molecules.

The spatial dimension examined in this chapter is still a little bit larger than such molecular clusters, but we are going to enter a regime where quantum size effects already start to kick in and where the local response approximation begins to fail [4–6]. A possible bridge between quantum and classical plasmonics has been introduced in [7], for example, where a quantum-corrected model is used to incorporate quantum-mechanical effects within a classical electrodynamic framework (also see the discussion in [8]). In this approach the junction between adjacent nanoparticles is modeled by means of a local dielectric response that additionally includes electron tunnelling and tunnelling resistivity at the gap. A variant of this method is discussed in [9], where nonlocal boundary conditions are employed within the MNPBEM toolbox. An improved and more advanced dielectric description that usually manifests itself in an additional spatial dependence seems to be a good starting point to overcome the difficulties for small particles and

---

[1]Within the *jellium model* one assumes a homogeneous electron gas, where the electrons interact with each other in presence of an uniform background of positive charges ("positive jelly", e.g. atomic cores in a solid). It is a simple model for the delocalized electrons in a metal, cf. Sect. 2.2. A more elaborate approach would be to incorporate also the influence of the atomic configuration in the plasmonic response, e.g. with ab initio calculations as presented in [3].

gap distances, without giving up the functionality and convenience of a classical field theory [10]. The appearance of such spatial dispersion effects on length scales comparable to the Thomas-Fermi screening length [11] establishes the term *nonlocal response*, in the sense that the dielectric function is not isotropic anymore and depends on the spatial position or the wave vector in Fourier space, respectively.

Within the local description, field enhancement and confinement are found to increase with decreasing gap distance between coupled nanoparticles, or when the radius of curvature becomes increasingly small at sharp tips and corners. Here the concept of a local response breaks down because of Landau damping, associated with the excitation of electron-hole pairs and corresponding processes involving wave vectors greater than ω/v_F, where ω as usual is the light frequency and v_F is again the Fermi velocity. The critical distance v_F/ω is then of the order of nanometers for visible and near-infrared light.

The spill-out of the valence electron density outside the metal also takes place over sub-nanometer distances (see Fig. 8.2), and gives rise to further nonlocal effects. In general, nonlocality leads to plasmon blueshifts, in comparison to a local description in these materials, as well as to plasmon broadening and significant reduction in the local field enhancement. These phenomena are thus detrimental for plasmonic applications in which confinement and field enhancement are critical. First experimental observations of nonlocal effects for small nanoparticles have recently been published [14, 15] (also see the discussion in [16, 17]), and the method of choice for these experiments was EELS. In particular, the resolution improvement of electron microscopes should allow systematic experimental studies of small gap regions and tiny particles.

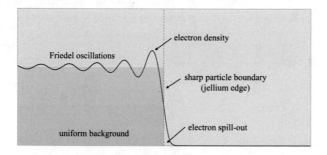

Fig. 8.2 Schematic representation of the electron charge density near a metal surface. Within Maxwell's theory we assume sharp and abrupt boundary conditions, but in reality there is a certain electron spill-out in close proximity to the surface. For further details and a self-consistent calculation of the electron density see e.g. [12] or [13]

8.1 Spatial Dependent Dielectric Function

One way to include spatial dispersion effects in the simulation of plasmonic nanoparticles is the modeling of a more sophisticated dielectric function [7, 18–25]. An approach that has already frequently been used in literature is the *hydrodynamic model*: The conduction band of the metal is treated as a classical electron plasma and an additional pressure term coming from the Pauli exclusion principle is introduced accounting for nonlocal effects. This leads to a wave vector dependent dielectric function of Drude form

Hydrodynamic model

$$\frac{\varepsilon(q,\omega)}{\varepsilon_0} = \varepsilon_\infty + \frac{\omega_p^2}{\beta^2 q^2 - \omega(\omega + \mathrm{i}\,\gamma_d)}, \tag{8.1}$$

where ε_∞ is again the ionic background for the corresponding material, ω_p is the bulk plasmon frequency, $\beta = \sqrt{3/5}\,v_F$ is the nonlocal parameter and γ_d is the phenomenological damping constant. For alkali metals this approach already recovers the expected blue shift and plasmon quenching (see Fig. 8.3), but nevertheless the results have to be treated with care since the hydrodynamical model describes the metal just as a compressed electron gas [20].

A more complex ansatz is to use a dielectric function derived by first principle calculations [26]. Based on first-order perturbation theory and the random-phase approximation, Lindhard showed that delocalized valence electron excitations pro-

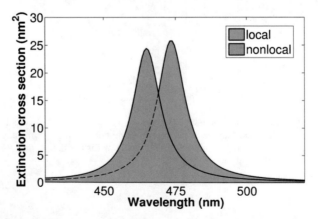

Fig. 8.3 Spectrum of a small gold sphere ($r = 2.5$ nm). The nonlocal result based on the hydrodynamical model shows a clear blue shift and plasmon quenching with respect to the local calculation

duce significant spatial dispersion. Later David Mermin self-consistently included electron-motion damping [27] and derived the final expression for a spatial dependent dielectric function [19].

References

1. M. Pelton, G.W. Bryant, *Introduction to Metal-Nanoparticle Plasmonics* (Wiley, Science Wise Publishing, Hoboken, New Jersey, 2013). ISBN 9781118060407
2. E. Townsend, G.W. Bryant, Plasmonic properties of metallic nanoparticles: the effects of size quantization. Nano Lett. **12**(1), 429–434 (2012).
3. P. Zhang, J. Feist, A. Rubio, P. García-González, F.J. García-Vidal, *Ab initio* nanoplasmonics: the impact of atomic structure. Phys. Rev. B **90**, 161407 (2014).
4. J. Zuloaga, E. Prodan, P. Nordlander, Quantum description of the plasmon resonances of a nanoparticle dimer. Nano Lett. **9**, 887–891 (2009).
5. J. Zuloaga, E. Prodan, P. Nordlander, Quantum plasmonics: optical properties and tunability of metallic nanorods. ACS Nano **4**, 2649 (2010).
6. K. Savage, M.M. Hawkeye, R. Esteban, A.G. Borisov, J. Aizpurua, J.J. Baumberg, Revealing the quantum regime in tunnelling plasmonics. Nature **491**, 574–577 (2012).
7. R. Esteban, A. Borisov, P. Nordlander, J. Aizpurua, Bridging quantum and classical plasmonics with a quantum-corrected model. Nat. Commun. **3**, 825 (2012).
8. R. Esteban, A. Zugarramurdi, P. Zhang, P. Nordlander, F.J. Garcia-Vidal, A.G. Borisov, J. Aizpurua, A classical treatment of optical tunneling in plasmonic gaps: extending the quantum corrected model to practical situations. Faraday Discuss. **178**, 151–183 (2015).
9. U. Hohenester, Quantum corrected model for plasmonic nanoparticles: a boundary element method approach. Phys. Rev. B **91**, 205436 (2015).
10. N.A. Mortensen, S. Raza, M. Wubs, T. Søndergaard, S.I. Bozhevolnyi, A generalized non-local optical response theory for plasmonic nanostructures. Nat. Commun. **5**(3809), 1 (2014).
11. N.W. Ashcroft, N.D. Mermin, *Festkörperphysik* (Oldenbourg, München, 2007). ISBN 978-3-48658273-4
12. N.D. Lang, W. Kohn, Theory of metal surfaces: charge density and surface energy. Phys. Rev. B **1**, 4555–4568 (1970).
13. A. Liebsch, Surface-plasmon dispersion and size dependence of Mie resonance: silver versus simple metals. Phys. Rev. B **48**, 11317–11328 (1993).
14. J.A. Scholl, A.L. Koh, J.A. Dionne, Quantum plasmon resonances of individual metallic nanoparticles. Nature **483**, 421 (2012).
15. S. Raza, N. Stenger, S. Kadkhodazadeh, S.V. Fischer, N. Kostesha, A.-P. Jauho, A. Burrows, M. Wubs, N.A. Mortensen, Blueshift of the surface plasmon resonance in silver nanoparticles studied with EELS. Nanophotonics **2**, 131 (2013).
16. H. Haberland, Looking from both sides. Nature **494**(7435), E1–E2 (2013).
17. T. Christensen, W. Yan, S. Raza, A.-P. Jauho, N.A. Mortensen, M. Wubs, Nonlocal response of metallic nanospheres probed by light, electrons, and atoms. ACS Nano **8**(2), 1745–1758 (2014).
18. R. Fuchs, F. Claro, Multipolar response of small metallic spheres: nonlocal theory. Phys. Rev. B **35**(8), 3722–3727 (1987).
19. F.J. Garcia de Abajo, Nonlocal effects in the plasmons of strongly interacting nanoparticles, dimers, and waveguides. J. Phys. Chem. C **112**, 17983–17987 (2008).
20. J. Aizpurua, A. Rivacoba, Nonlocal effects in the plasmons of nanowires and nanocavities excited by fast electron beams. Phys. Rev. B **78**(3), 035404 (2008).
21. J.M. McMahon, S.K. Gray, G.C. Schatz, Nonlocal optical response of metal nanostructures with arbitrary shape. Phys. Rev. Lett. **103**, 097403 (2009).
22. C. David, F.J. Garcia de Abajo, Spatial nonlocality in the optical response of metal nanoparticles. J. Phys. Chem. C **115**, 19470 (2011).
23. A.I. Fernández-Domínguez, A. Wiener, F.J. García-Vidal, S.A. Maier, J.B. Pendry, Transformation-optics description of nonlocal effects in plasmonic nanostructures. Phys. Rev. Lett. **108**, 106802 (2012).

24. W. Yan, N.A. Mortensen, M. Wubs, Green's function surface-integral method for nonlocal response of plasmonic nanowires in arbitrary dielectric environments. Phys. Rev. B **88**, 155414 (2013).
25. L. Stella, P. Zhang, F.J. García-Vidal, A. Rubio, P. García-González, Performance of nonlocal optics when applied to plasmonic nanostructures. J. Phys. Chem. C **117**, 8941–8949 (2013).
26. F.J. García de Abajo, Optical excitations in electron microscopy. Rev. Mod. Phys. **82**, 209 (2010).
27. N.D. Mermin, Lindhard dielectric function in the relaxation-time approximation. Phys. Rev. B **1**, 2362–2363 (1970).

Chapter 9
Metamaterials

I was invisible, and I was only just beginning to realise the
extraordinary advantage my invisibility gave me. My head was
already teeming with plans of all the wild and wonderful things
I had now impunity to do.

H. G. WELLS, The Invisible Man

The refractive index n corresponds to the factor by which the speed and wavelength of any radiation is reduced, when it propagates in an optical medium rather than in vacuum. Hence it describes a ratio and is therefore a dimensionless number. A value of $n = 1.5$, for example, states that a light wave travels 1.5 times faster in vacuum than it does in the corresponding medium (glass in this case). Since we also know the laws of relativity and its keystone, the absolute value and constancy of the speed of light, it is only reasonable that n has to be a positive number greater than unity for all optical materials in our universe. If we investigate an absorbing medium like a metal for example, we have seen that the absorption can be described by an imaginary refractive index which results from

$$n(\omega) = \frac{c_{\text{vac}}}{c_{\text{med}}} = \sqrt{\frac{\varepsilon(\omega)\mu(\omega)}{\varepsilon_0\mu_0}} = \sqrt{\varepsilon_r(\omega)\mu_r(\omega)}. \qquad (9.1)$$

So now we obtain a complex number where the imaginary part accounts for absorption, but still, we would assume that the real part of n has to be greater than 1. If we now lock the door, draw the curtains and forget about relativity just for a very brief moment–nothing forbids the square root to give a positive *and* a negative result. What does this mean? Are values of n less than 1 or even negative refractive indices possible? First of all there is of course no conflict with relativity (the caution with door and curtains was uncalled-for): The refractive index only measures the corresponding *phase velocity* of the radiation, which does not carry information and it can be shown [1], that signals with a phase velocity greater than c still move with a *group velocity* less than (or equal to) c. That's that, but what does a negative sign in Eq. (9.1) imply? It means that we could shed the limitations of conventional optics and do extraordinary things, it allows the composition of so-called *superlenses* [2, 3] or the formation of exotic new materials that behave very different from what we know in real life. We call this new kind of negative

© Springer International Publishing Switzerland 2016
A. Trügler, *Optical Properties of Metallic Nanoparticles*, Springer Series
in Materials Science 232, DOI 10.1007/978-3-319-25074-8_9

Fig. 9.1 Ray tracing picture of a glass of water with a negative refractive index by E. Schrempp [10], also see [11] and [12]. The difference in the optical density of air and "normal" water (*left*) causes a straw in the glass to seem to be shifted at the interface and slightly enlarged inside the liquid. In "negative-index water" (*right*), the straw would seem to continue in "the wrong direction". Figure reprinted by permission from Macmillan Publishers Ltd: Nature [10], © 2008

refraction materials *metamaterials* [4–9, 41] and once we are able to control ε and μ in Eq. (9.1) we can manipulate electromagnetic waves in very unusual ways (see Fig. 9.1) and even make things invisible with so-called *cloaking devices*.

Usual matter is formed by crystal lattices and structured arrangements of atoms. A metamaterial follows this requirement, it is a structured arrangement of artificial elements, designed to achieve a certain dielectric behavior and consequently advantageous and unusual electromagnetic properties (see e.g. [6]). The light effectively averages over these periodically structured "photonic atoms" [1], which of course is very similar to our discussion of the dielectric response of a metal in Chap. 2. Since we want to avoid diffraction, these artificial atoms have to be small compared to the wavelength of the impinging radiation. Figure 9.2 shows a ε_r-μ_r diagram [13] that divides space into four separate regions. Until now we have moved solely along the dotted horizontal line: For optical frequencies we have $\mu_r = \mu/\mu_0 \approx 1$, the negative ε for metals on the left hand side allows the formation of evanescent waves and common transparent dielectrics with $\varepsilon_r = \varepsilon/\varepsilon_0 \geq 1$ are situated on the right upper quarter. Only if both, ε and μ become negative, we end up with a negative refractive index. This can be easily seen if we recast Eq. (9.1) into

$$n = \sqrt{|\varepsilon_r||\mu_r|}\, e^{\frac{i}{2}(\varphi_{\varepsilon_r}+\varphi_{\mu_r})}, \tag{9.2}$$

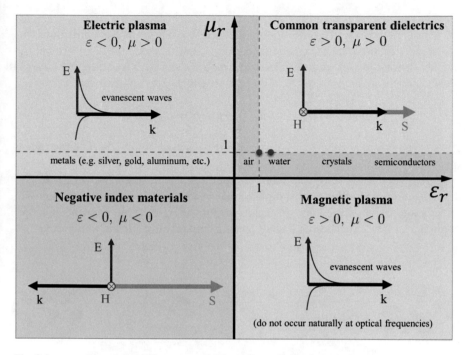

Fig. 9.2 ε_r-μ_r diagram for different materials, see discussion in the main text

where the angles φ are the arguments of the corresponding complex variable and have to be determined in the correct sector of the complex plane,

$$\varphi_{\varepsilon_r} = \arctan\left[\frac{\Im m(\varepsilon_r)}{\Re e(\varepsilon_r)}\right], \qquad \varphi_{\mu_r} = \arctan\left[\frac{\Im m(\mu_r)}{\Re e(\mu_r)}\right]. \qquad (9.3)$$

For $\Re e(\varepsilon) < 0$, $\Re e(\mu) > 0$ (electric plasma) or $\Re e(\varepsilon) > 0$, $\Re e(\mu) < 0$ (magnetic plasma) we still obtain a positive n, only if both material parameters become negative, we obtain negative refraction.[1] It should also be noted here that with this regard the term metamaterial does not automatically implicate a negative n. The artificial magnetic structures in the lower right quarter of Fig. 9.2 are also called metamaterials and especially when it comes to cloaking devices usually a positive but spatially varying refractive index is required.

One of the first strange characteristics of negative refraction is also shown in the vector plot inset of Fig. 9.2: The negative sign reverses the direction of the energy flux $S = E \times H$. For common dielectrics the Poynting vector S points into the

---

[1] Any negative-index material must be strongly dispersive, i.e. there must also exist frequency ranges with a positive refractive index, because otherwise the energy density integrated over all frequencies would be negative [14].

propagation direction k and we obtain so-called right-handed materials, whereas for negative index materials the energy flux and k point in opposite directions. In this case k and the electromagnetic fields form a left-handed material. This can be quickly verified [13] if we have a look at Ampère's and Faraday's law (3.3). For a plane monochromatic wave $e^{ik \cdot r - i\omega t}$ and the absence of any sources we get

$$k \times H = -\omega\varepsilon E, \qquad k \times E = \omega\mu H. \tag{9.4}$$

If we change from positive to negative ε and μ we switch from right- to left-handed materials.

Before we discuss some selected topics out of this vast research field, let us get a quick impression of what three dimensional metamaterials actually look like. We already know that we require small and periodically structured elements–a more explicit overview over several 3D photonic metamaterials is plotted in Fig. 9.3.

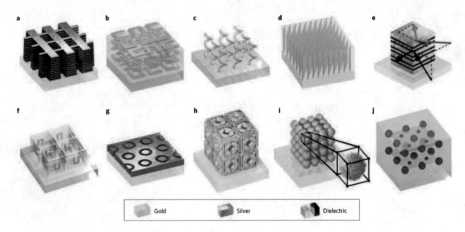

Fig. 9.3 3D photonic-metamaterial structures reprinted by permission from Macmillan Publishers Ltd: Nature Photonics [15], © 2011. References to the corresponding structures and further information can be also found in [15]. (**a**) Double-fishnet negative-index metamaterial. (**b**) 'Stereo' or chiral metamaterial fabricated through stacked e-beam lithography. (**c**) Chiral metamaterial made using direct-laser writing. (**d**) Hyperbolic metamaterial. (**e**) Metal-dielectric layered metamaterial composed of coupled plasmonic waveguides. (**f**) Split-ring resonators oriented in all three dimensions, fabricated using membrane projection lithography. (**g**) Wide-angle visible negative-index metamaterial based on a coaxial design. (**h**) Connected cubic-symmetry negative-index metamaterial. (**i**) Metal cluster-of-clusters visible-frequency magnetic metamaterial made using self-assembly. (**j**) All-dielectric negative-index metamaterial

9.1 The Veselago Lens and Superresolution

In Chap. 5 we briefly discussed the limits of conventional optics–no matter how perfectly we polish a lens, the achievable resolution in the far-field region will always be of the order of the wavelength. By probing the electromagnetic field very close to the investigated specimen (e.g. through SNOM experiments) we have seen that we can bypass the limitations and extract information beyond the classical diffraction limit. But in such near-field measurements we have to bring a probe, a certain antenna in very close proximity to the nanoobject and the final resolution is determined by the size of this probe.

In 1967 the Russian physicist Victor Veselago proposed a new type of lens based on metamaterials [13] and in 2000 John Pendry showed that these new lenses allow (almost[2]) perfect reconstruction of an image beyond practical limitations of apertures or lenses [18], since they cancel the decay of evanescent waves. With this new *superlenses* both propagating and evanescent waves contribute to the resolution of the image [18]. Pendry discussed the reconstruction of an image in the near-field zone, but such a lens also builds a perfect image of a three-dimensional nanoobject in the far-field region [2].

A Veselago lens is quite different from a typical optical lens; there is no curvature, no optical axis, no magnification and parallel light beams cannot be focused (see Fig. 9.4). Basically it is just a vertical, parallel-sided slab, but with the very important feature of a negative refractive index. This has very strange consequences

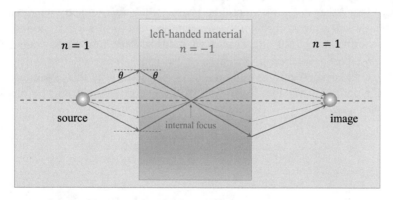

Fig. 9.4 With negative refraction, the light waves from an external source are bent beyond the normal to the interface (*dotted arrows*) at both interfaces of a Veselago lens and produce a perfect image of the object [19]

---

[2]In [16] the authors discovered a fundamental limitation on the ultimate spatial resolution of the perfect lens as a result from spatial dispersion (nonlocality) of the dielectric response. The resolution of the lens will also generally be reduced if the slab material is lossy, see e.g. [17] for a summary to this topic.

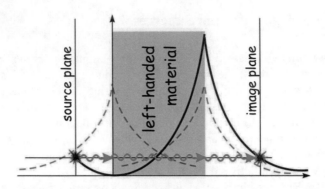

Fig. 9.5 A slab of an ideal left-handed material represents two coupled plasmon-polariton resonators. It is transparent for propagating fields (*green wavy line*) and plasmon polaritons are always in resonance with evanescent fields from the source. The two coupled surface modes at each side of the slab (*dashed red lines*) [21] then resonantly enhance the evanescent field from the source (*solid red line*) and as a result, both propagating and evanescent fields form an exact copy of the source field in the focal point. Reprinted with permission from [3]. © 2008 by the American Physical Society

like the reversal of equivalents of Snell's law, the Doppler shift, or Cherenkov radiation [2, 13]. A conventional lens applies a phase correction to each of the impinging Fourier components of the electromagnetic field so that at some distance beyond the lens the fields reassemble to a focus. Since the evanescent fields only decay in amplitude but not in phase, the requirement for a corresponding focus is to amplify them rather than to correct their phase [18]. This is exactly what happens inside the Veselago lens, through the resonant excitation of surface modes at the boundary of the slab [4, 20], the evanescent fields become amplified as depicted in Fig. 9.5.

Because the evanescent waves do not carry any net energy flux, the energy can never be amplified; only the distribution of the energy or field will be modified across space [2].

9.2 Artificial Magnetic Atoms

Let us now briefly explore the lower right quarter of Fig. 9.2. We know from magnetostatics that a circulating ring current of a microscopic coil yields a certain magnetic-dipole moment given by the product of current and area of the coil [1]. This dipole moment can be increased if we combine the coil with a plate capacitor which leads to a magnetic resonance. Thus, a popular design for *magnetic atoms* is to mimic an ordinary LC-circuit, consisting of a plate capacitor with capacitance C and a magnetic coil with inductance L, on a scale much smaller than the relevant wavelength of light [1], see Fig. 9.6. The LC-circuit in this sense is an analogue of a tuning fork, where the oscillation of the fork corresponds to the energy oscillating at the circuit's resonance frequency $\omega_{LC} = 1/\sqrt{LC}$.

In Sect. 2.2 we have seen that the dielectric response of a metal (upper left quarter in Fig. 9.2) can be approximated by the relation

$$\frac{\varepsilon_d}{\varepsilon_0} \approx 1 - \frac{\omega_p^2}{\omega^2}. \tag{9.5}$$

The effective magnetic permeability for an array of split-ring resonators can be brought into an analogue form

Magnetic permeability of split-ring resonators

$$\frac{\mu}{\mu_0} \approx 1 - \frac{\omega_{mp}^2}{\omega^2}, \tag{9.6}$$

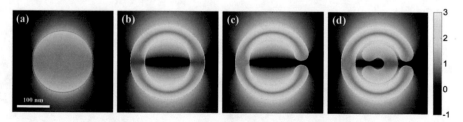

Fig. 9.6 Principle of magnetic atoms as LC-circuits. The panels show the magnetic field for each geometry for a plane wave excitation polarized along the x-axis. (**a**) The magnetic response for a regular gold nanodisk is negligible. (**b**) If we change from a disk to a ring design, the loop current induces a magnetic field (or vice versa) and the magnetic response becomes somewhat enhanced. (**c**) Cutting the geometry into a split-ring [22–24] introduces a magnetic resonance analogous to an LC-circuit. (**d**) The magnetic resonance can be enhanced if the split-ring is doubled

with a corresponding magnetic plasma frequency ω_{mp}. The derivation for this equation can be found in [24], for example, also see the discussion in [1]. Based on the resonance behavior the split-ring design is essential to obtain $\Re e(\mu) < 0$ [25], but as plotted in Fig. 9.7 several variants of the base design are possible.

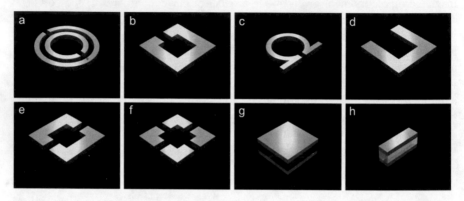

Fig. 9.7 Typical designs of magnetic atoms for photonics (a-h). Reprinted from [1], © 2007, with permission from Elsevier

9.3 Making Things Invisible

The basic principle behind metamaterials is to mold the flow of light with the help
of resonant periodic structures that are considerably smaller than the corresponding
photon wavelength. Although with negative refraction several principles of optics
are turned upside down, in the end everything still comes out of Maxwell's
equations. When you think about molding the propagation of electromagnetic waves
a question that will arise sooner or later is how do Maxwell's equations look like
in curved space? Will the laws of Gauss, Ampère and Faraday still be the same
if we leave the safe harbor of Euclidean space where they have been developed?
John Pendry and coworkers gave an answer to this interesting problem [26, 27]
and thereby built the foundation of *transformation optics* [28–32]. If we perform
a coordinate transformation and change from Cartesian coordinates x, y, z to an
arbitrary coordinate system u, v, w (see Fig. 9.8), Maxwell's equations have exactly
the same form, only the permittivity ε and permeability μ have to be scaled by a
common factor

Transformation optics

$$\varepsilon'_u = \varepsilon_u \frac{Q_u Q_v Q_w}{Q_u^2}, \qquad \mu'_u = \mu_u \frac{Q_u Q_v Q_w}{Q_u^2}, \qquad \text{etc.,} \qquad (9.7)$$

with the Jacobian transformation

$$Q_u^2 = \left(\frac{\partial x}{\partial u}\right)^2 + \left(\frac{\partial y}{\partial u}\right)^2 + \left(\frac{\partial y}{\partial u}\right)^2, \qquad (9.8a)$$

$$Q_v^2 = \left(\frac{\partial x}{\partial v}\right)^2 + \left(\frac{\partial y}{\partial v}\right)^2 + \left(\frac{\partial y}{\partial v}\right)^2, \qquad (9.8b)$$

$$Q_w^2 = \left(\frac{\partial x}{\partial w}\right)^2 + \left(\frac{\partial y}{\partial w}\right)^2 + \left(\frac{\partial y}{\partial w}\right)^2. \qquad (9.8c)$$

The consequences are remarkable, instead of bending space we only need to adjust
the response functions ε and μ to obtain the same effect. Using the possibilities
that the design of metamaterials provide, we are able to redirect electromagnetic
fields at will [26], to steer light around objects and hence make them seem
invisible. In Fig. 9.9 different types of invisibility cloaks that use materials with
a spatially changing refractive index are shown. The first practical realization of
such a cloak has been reported in [34], where artificially structured metamaterials
operating over a band of microwave frequencies have been used to conceal a
copper cylinder. Over the last few years the operating frequencies of metamaterials
successively approached toward the visible spectrum, see the summary in Fig. 9.10.

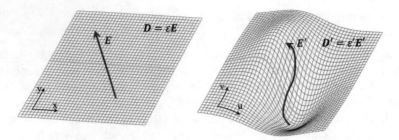

Fig. 9.8 Schematic representation of the transformation from a flat Euclidean space to a distorted mesh with curvilinear coordinates. A *straight light line* (*yellow*) in the *left panel* is bent in the new system, likewise the electric field vector *E*

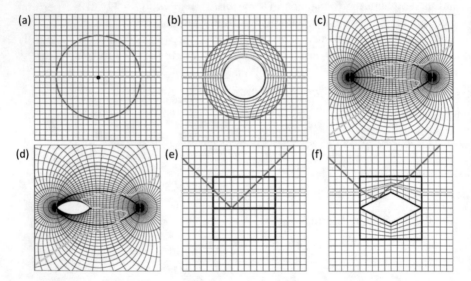

Fig. 9.9 Light propagates in a straight line in free space (**a**). By carrying out so-called *push-forward mapping* [26] to expand a point (*red dot* in **a**) into a spherical hole (*red circle* in **b**), light will be guided smoothly around the hole if the material inside the compressed region is prescribed according to Eq. (9.7). Parts (**c**) and (**d**) illustrate *conformal-mapping* [28] (based on ray optics). Using non-Euclidean geometry, conformal mapping can be devised so that the light rays (*yellow lines*) from any direction will never reach a line (or an infinitesimally thin plate, *red line* in **c**) or a closed region (*red eye* in **d**). The basic idea of a *carpet cloak* is shown in (**e**) and (**f**). Further discussion see [33] and [29]. Reprinted by permission from Macmillan Publishers Ltd: Nature Materials [33], © 2010

The dream of a cloak of invisibility is as old as mankind and with the development of metamaterials we are one step closer to the concept moving from fiction to reality, see Fig. 9.11. Unfortunately the fans of H. G. Wells, Harry Potter, or the Nibelungs will still have to be patient, because with this approach we achieve perfect

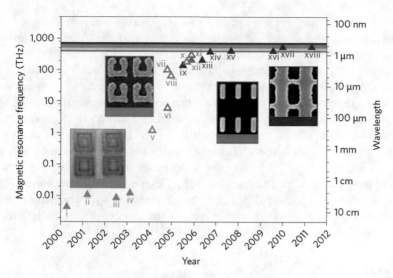

Fig. 9.10 Progress in metamaterial operating frequency over the past decade. The operating frequency of metamaterials with negative magnetic permeability μ (*empty triangles*) and negative index of refraction n (*solid triangles*) is shown on a logarithmic scale ranging from microwave to visible wavelengths. *Orange*: structures based on double split-ring resonators; *green*: U-shaped split-ring resonators; *blue*: metallic cut-wire pairs; *red*: negative-index double fishnet structures. The four insets show optical or electron micrographs of the four types of structure. Reprinted by permission from Macmillan Publishers Ltd: Nature Photonics [15], © 2011. Also see [35]

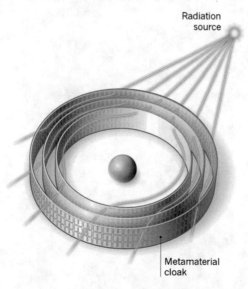

Fig. 9.11 An invisibility cloak made of a negative index metamaterial bends radiation around an object inside it. Reprinted by permission from Macmillan Publishers Ltd: Nature [36], © 2013

invisibility only in one single frequency of light.[3] This is inevitable because the required effective permittivity and permeability can only be realized by resonances in the building blocks of the metamaterials, and causality requirements dictate that perfect invisibility cannot have wide bandwidth [33]. The fictional invisibility cloaks also allow their possessors to see the outside world while they themselves are concealed behind the cloak. By contrast if a metamaterial cloaking device encircles an object, electromagnetic waves are guided around it and no light can reach the object making it impossible to see the outside world. Nevertheless the possibility to create materials with a spatially varying index has impressive potentials and transformation optics has evolved into a powerful tool for designing a wide variety of new optical effects and devices [33]. Computational metamaterials [37], applications beyond optics [38] or again the combination with graphene [39] are currently being explored, along with more curious aspects of metamaterials like electromagnetic wormholes [40].

---

[3]Recently some strategies have been developed to overcome this narrow bandwidth constraint, mostly by sacrificing some degree of invisibility in return for a broader bandwidth [33].

References

1. K. Busch, G. von Freymann, S. Linden, S. Mingaleev, L. Tkeshelashvili, M. Wegener, Periodic nanostructures for photonics. Phys. Rep. **444**, 101–202 (2007).
2. X. Zhang, Z. Liu, Superlenses to overcome the diffraction limit. Nat. Mater. **7**(6), 435–441 (2008).
3. K.Y. Bliokh, Y.P. Bliokh, V. Freilikher, S. Savel'ev, F. Nori, *Colloquium*: unusual resonators: plasmonics, metamaterials, and random media. Rev. Mod. Phys. **80**, 1201–1213 (2008).
4. D.R. Smith, J.B. Pendry, M.C.K. Wiltshire, Metamaterials and negative refractive index. Science **305**(5685), 788–792 (2004).
5. H.J. Lezec, J.A. Dionne, H.A. Atwater, Negative refraction at visible frequencies. Science **316**, 5823 (2007).
6. A.K. Sarychev, V.M. Shalaev, *Electrodynamics of Metamaterials* (World Scientific, Singapore/Hackensack, 2007). ISBN 9789810242459
7. W. Cai, V. Shalaev, *Optical Metamaterials: Fundamentals and Applications* (Springer, Berlin, 2010). ISBN 978-1-4419-1150-6
8. N.I. Zheludev, A roadmap for metamaterials. Opt. Photonics News **22**, 30–35 (2011).
9. F. Monticone, A. Alú, Metamaterials and plasmonics: from nanoparticles to nanoantenna arrays, metasurfaces, and metamaterials. Chin. Phys. B **23**(4), 047809 (2014).
10. O. Hess, Optics: farewell to flatland. Nature **455**(7211), 299–300 (2008).
11. G. Dolling, M. Wegener, S. Linden, C. Hormann, Photorealistic images of objects in effective negative-index materials. Opt. Exp. **14**(5), 1842–1849 (2006).
12. D. Schurig, J.B. Pendry, D.R. Smith, Calculation of material properties and ray tracing in transformation media. Opt. Exp. **14**(21), 9794–9804 (2006).
13. V.G. Veselago, The electrodynamics of substances with simultaneously negative values of ε and μ. Sov. Phys. Usp. **10**, 509 (1968).
14. L. Novotny, B. Hecht, *Principles of Nano-Optics*, 2nd edn. (Cambridge University Press, Cambridge, 2012). ISBN 978-1107005464
15. C.M. Soukoulis, M. Wegener, Past achievements and future challenges in the development of three-dimensional photonic metamaterials. Nat. Photonics **5**, 523–530 (2011).
16. I.A. Larkin, M.I. Stockman, Imperfect perfect lens. Nano Lett. **5**, 339 (2005).
17. R.E. Collin, Frequency dispersion limits resolution in veselago lens. Prog. Electromagn. Res. B **19**, 233–261 (2010).
18. J.B. Pendry, Negative refraction makes a perfect lens. Phys. Rev. Lett. **85**, 3966–3969 (2000).
19. R. Sambles, Nano-optics: gold loses its lustre. Nat. Photonics **438**(7066), 295–296 (2005).
20. P. Andrew, W.L. Barnes, Energy transfer across a metal film mediated by surface plasmon polaritons. Science **306**(5698), 1002–1005 (2004).
21. W.L. Barnes, A. Dereux, T.W. Ebbesen, Surface plasmon subwavelength optics. Nature **424**(6950), 824–830 (2003).
22. W.N. Hardy, L.A. Whitehead, Split-ring resonator for use in magnetic resonance from 200-2000 MHz. Rev. Sci. Instrum. **52**(2), 213–216 (1981).
23. J.B. Pendry, A.J. Holden, W.J. Stewart, I. Youngs, Extremely low frequency plasmons in metallic mesostructures. Phys. Rev. Lett. **76**, 4773–4776 (1996).
24. J.B. Pendry, A.J. Holden, D.J. Robbins, W.J. Stewart, Magnetism from conductors and enhanced nonlinear phenomena. IEEE Trans. Microwave Theory Tech. **47** (1999) 2075–2084.
25. S. Linden, C. Enkrich, M. Wegener, J. Zhou, T. Koschny, C.M. Soukoulis, Magnetic response of metamaterials at 100 Terahertz. Science **306**(5700), 1351–1353 (2004).

26. J.B. Pendry, D. Schurig, D.R. Smith, Controlling electromagnetic fields. Science **312**(5781), 1780–1782 (2006).
27. A.J. Ward, J.B. Pendry, Refraction and geometry in Maxwell's equations. J. Mod. Opt. **43**(4), 773–793 (1996).
28. U. Leonhardt, Optical conformal mapping. Science **312**(5781), 1777–1780 (2006).
29. U. Leonhardt, T. Tyc, Broadband invisibility by non-Euclidean cloaking. Science **323**(5910), 110–112 (2009).
30. A. Nicolet, F. Zolla, C. Geuzaine, Transformation optics, generalized cloaking and superlenses. IEEE Trans. Magn. **46**(8), 2975–2981 (2010).
31. J.B. Pendry, A. Aubry, D.R. Smith, S.A. Maier, Transformation optics and subwavelength control of light. Science **337**(6094), 549–552 (2012).
32. A.I. Fernández-Domínguez, A. Wiener, F.J. García-Vidal, S.A. Maier, J.B. Pendry, Transformation-optics description of nonlocal effects in plasmonic nanostructures. Phys. Rev. Lett. **108**, 106802 (2012).
33. H. Chen, C.T. Chan, P. Sheng, Transformation optics and metamaterials. Nat. Mater. **9**(5), 387–396 (2010).
34. D. Schurig, J.J. Mock, B.J. Justice, S.A. Cummer, J.B. Pendry, A.F. Starr, D.R. Smith, Metamaterial electromagnetic cloak at microwave frequencies. Science **314**(5801), 977–980 (2006).
35. C. M. Soukoulis, S. Linden, M. Wegener, Negative refractive index at optical wavelengths. Science **315**(5808), 47–49 (2007) .
36. L. Billings, Exotic optics: metamaterial world. Nature **500**(7461), 138–140 (2013).
37. A. Silva, F. Monticone, G. Castaldi, V. Galdi, A. Alú, N. Engheta, Performing mathematical operations with metamaterials. Science **343**(6167), 160–163 (2014).
38. M. Wegener, Metamaterials beyond optics. Science **342**(6161), 939–940 (2013).
39. A. Vakil, N. Engheta, Transformation optics using graphene. Science **332**(6035), 1291–1294 (2011).
40. A. Greenleaf, Y. Kurylev, M. Lassas, G. Uhlmann, Electromagnetic wormholes and virtual magnetic monopoles from metamaterials. Phys. Rev. Lett. **99**, 183901 (2007).
41. U. Leonhardt, T. Philbin, Geometry and Light: The Science of Invisibility. Dover Books on Physics (Dover, New York, 2010). http://store.doverpublications.com/0486476936.html [ISBN 978-0486476933]

Chapter 10
Outlook

The world is a thing of utter inordinate complexity and richness and strangeness that is absolutely awesome.

DOUGLAS ADAMS

We have discussed many aspects of the optical properties of metallic nanoparticles, ranging from the tunability of the plasmonic resonance to the ultrafast dynamics of the light-matter interaction, from the possibility to map the plasmonic fields with EELS and tomography methods to nonlocal plasmonics, from negative refraction to invisibility cloaks. There is one conspicuous thing about all these applications in plasmonics: On one hand we require metals to obtain the desired concentration of electromagnetic fields at the nanoscale, but on the other hand it is these metals that limit the possibilities because of their lossy nature. It has been mentioned several times in this book that the combination with other materials often allows us to overcome some of the drawbacks of metals and hence the search for novel materials with lower loss in plasmonics is on the agenda of many research groups all over the world. In [1] Jacob Khurgin provokingly states that

> This search, however, should not follow today's pattern of rapid-fire testing of all the well-known conductors, doped semiconductors, or popular materials *du jour* (graphene, MoS$_2$, and whatever comes next), in hope of a miracle, but instead should be a well thought-out and concerted effort by condensed-matter theorists, chemists and growth specialists to synthesize man-made negative permittivity materials with reduced loss.

It is truly amazing what has already been achieved in plasmonics and I'm sure that such joint efforts hold great promise for the future direction of the research field. This book started with a vision of Richard Feynman about the fantastic possibilities down at the bottom [2] and he was absolutely right about it:

> As soon as I mention [manipulating and controlling things on a small scale], people tell me about miniaturization, and how far it has progressed today. They tell me about electric motors that are the size of the nail on your small finger. And there is a device on the market, they tell me, by which you can write the Lord's Prayer on the head of a pin. But that's nothing; that's the most primitive, halting step in the direction I intend to discuss. It is a staggeringly small world that is below. In the year 2000, when they look back at this age, they will wonder why it was not until the year 1960 that anybody began seriously to move in this direction. [...] I am not inventing anti-gravity, which is possible someday only if the

© Springer International Publishing Switzerland 2016
A. Trügler, *Optical Properties of Metallic Nanoparticles*, Springer Series in Materials Science 232, DOI 10.1007/978-3-319-25074-8_10

laws are not what we think. I am telling you what could be done if the laws *are* what we think.

In the end the key to success in any kind of fundamental research is to find the unexpected and forcing science into the corset of forecasted directions will not work. Thus the prediction of future perspectives in any kind of research field is inherently risky. Nevertheless because of the highly diverse nature of plasmonics we can assume an exciting future and high activity in this field as discussed in [3, 4], where, for example, the merging of plasmonics with quantum systems, active plasmonic devices, biochemical applications or electronic transport are named as future trends.

There are still many open questions to explore and I hope that you, gentle reader, are now at the end of this book more enthusiastic than ever to contribute something to the scientific progress of this fascinating field of research.

References

1. J. B. Khurgin, How to deal with the loss in plasmonics and metamaterials. Nat. Nanotechnol. **10**(1), 2–6 (2015).
2. R. Feynman, There is plenty of room at the bottom (talk transcript). Caltech Eng. Sci. **23**(5), 22–36 (1960).
3. N. Halas, Plasmonics: an emerging field fostered by nano letters. Nano Lett. **10**, 3816–3822 (2010).
4. Perspective on plasmonics. Nat. Photonics **6**(11), 714–715 (2012).

Appendix A
Utilities

A.1 Conversion Between nm and eV

The energy of a photon is given by

$$E = \hbar \omega = \hbar \, 2\pi v = \hbar \, 2\pi \frac{c}{\lambda}. \tag{A.1}$$

The SI-values for the involved constants can be found in [1], for example, and are given by

$$\hbar = 6.58211928 \times 10^{-16} \quad [\text{eV s}], \tag{A.2}$$

$$c = 2.99792458 \times 10^{17} \quad [\text{nm/s}]. \tag{A.3}$$

From Eq. (A.1) then follows[1]

Conversion from eV to nm

$$[\text{eV}] = {}^{1239.84}/_{[\text{nm}]} \tag{A.4}$$

The values of the visible spectrum of light in eV and nm are shown in Fig. A.1.

---

[1] In [2] also the value 8065.48 $[\text{cm}^{-1}/\text{ev}]$ is published.

© Springer International Publishing Switzerland 2016
A. Trügler, *Optical Properties of Metallic Nanoparticles*, Springer Series
in Materials Science 232, DOI 10.1007/978-3-319-25074-8

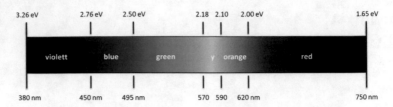

Fig. A.1 Energy values of the visible spectrum in nm and eV, the approximated color ranges have been adopted from [3]. The range from 750 to 3000 nm (1.65–0.41 eV) is called near-infrared, the region from 10 to 380 nm (123.98–3.26 eV) ultraviolet [3]

A.2   Conversion Between FWHM and Decay Time

By following [4], let us investigate the damped oscillation of an electric field $E(t)$. The exponential decay shall be described by a time constant τ, the eigenfrequency by ω_0. From the Fourier transform of $E(t)$

$$E(t) = E_0 e^{-i\omega_0 t} e^{-t/\tau}, \tag{A.5}$$

$$\mathcal{F}\{E(t)\} = \frac{E_0}{\sqrt{2\pi}} \frac{1/\tau}{(\omega - \omega_0)^2 + (1/\tau)^2}, \tag{A.6}$$

follows a so called *Lorentz-lineshape* (see Fig. A.2), which is characterized by its center position ω_0 and the full width at half maximum (FWHM) $\Delta\omega$. In units of energy the FWHM is given by $\Gamma = \hbar\Delta\omega$. By determining the extrema it follows that half of the maximum for Eq. (A.6) is reached at $(\omega - \omega_0) = 1/\tau$, which directly leads to the connection of the FWHM with the decay time: $\Gamma = 2\hbar/\tau$. With Eqs. (A.2) and (A.4) we finally obtain

---

Conversion between FWHM and decay time

$$\tau\ [\text{fs}] = \frac{1.316424}{\Gamma\ [\text{eV}]} \qquad \text{or} \qquad \tau\ [\text{fs}] = \kappa \cdot 1.061767 \times 10^{-3}, \tag{A.7}$$

---

where we have used the abbreviation

$$\kappa = \left(\frac{1}{\lambda_1} - \frac{1}{\lambda_2}\right)^{-1} = \frac{\lambda_1\lambda_2}{\lambda_2 - \lambda_1}.$$

Here λ_1 and λ_2 correspond to the left and right limits of the FWHM (see Fig. 2.31), respectively.

The same relation also holds for a driven harmonic oscillator, see [4] for further details. Note that the oscillator energy (amplitude squared) decays twice as fast as the oscillation amplitude!

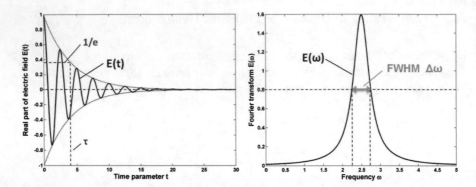

Fig. A.2 Lorentzian curve [*right panel*, Eq. (A.6)] obtained by the Fourier transform of a damped oscillator in the time domain [*left panel*, Eq. (A.5)] [4]. Chosen parameters: Damping time $\tau = 4$, frequency $\omega_0 = 2.5$. The FWHM is $\Delta\omega = 0.5$, which reproduces the damping time: $2/\Delta\omega = 4$

A.3   Derivation of Retarded Surface Charges and Currents

The continuity of the tangential magnetic field follows as

$$H_1 h_1 - H_2 h_2 - i\omega\hat{n}(G_1\mu_1\varepsilon_1\sigma_1 - G_2\mu_2\varepsilon_2\sigma_2) =$$
$$\Delta A'_{\text{ext}} + ik\hat{n}(\mu_1\varepsilon_1\phi_1 - \mu_2\varepsilon_2\phi_2), \qquad \text{(A.8)}$$

where $H_{1,2} = F \pm 2\pi\mathbb{1}$. The continuity of the normal electric displacement gives

$$H_1\varepsilon_1\sigma_1 - H_2\varepsilon_2\sigma_2 - i\omega\hat{n}\cdot(G_1\varepsilon_1 h_1 - G_1\varepsilon_1 h_2) =$$
$$\varepsilon_1(i\omega\hat{n}\cdot A_1^{\text{ext}} - \phi'_1) - \varepsilon_2(i\omega\hat{n}\cdot A_2^{\text{ext}} - \phi'_2). \qquad \text{(A.9)}$$

We now insert Eq. (3.74) into Eqs. (A.8) and (A.9) and use

$$H_1 h_1 = H_1\ G_1^{-1}(G_2 h_2 + \Delta A_{\text{ext}}) = \Sigma_1(G_2 h_2 + \Delta A_{\text{ext}}), \qquad \text{(A.10)}$$

$$H_2 h_2 = H_2\ G_2^{-1}(G_2 h_2\qquad\quad) = \Sigma_2(G_2 h_2), \qquad \text{(A.11)}$$

where $\Sigma_{1,2} = H_{1,2}\,G_{1,2}^{-1}$. Furthermore we obtain $H_1\mu_1\varepsilon_1\sigma_1 = \Sigma_1 G_1 L_1\sigma_1$, where $L_{1,2} = G_{1,2}\mu_{1,2}\varepsilon_{1,2}G_{1,2}^{-1}$. After some more algebra we obtain

$$(\Sigma_1\quad - \Sigma_2\ )G_2 h_2 - i\omega\hat{n}(L_1 - L_2)G_2\sigma_2 =$$
$$\Delta A'_{\text{ext}} + i\omega\hat{n}(\mu_1\varepsilon_1\phi_1 - \mu_2\varepsilon_2\phi_2) - \Sigma_1\Delta A_{\text{ext}} + i\omega\hat{n}L_1\Delta\phi_{\text{ext}},$$

$$(\Sigma_1 L_1 - \Sigma_2 L_2)G_2\sigma_2 - i\omega\hat{n}(L_1 - L_2)G_2 h_2 =$$
$$D'_e\ -\ \Sigma_1 L_1\Delta\phi_{\text{ext}}\ +\ i\omega\hat{n}\ \cdot\ L_1\Delta A_{\text{ext}}.$$

For a compact notation and in agreement with [5] we introduce the following abbreviations[2]:

$$D_e = D'_e - \Sigma_1 L_1\Delta\phi_{\text{ext}} + i\omega\hat{n}\cdot L_1\Delta A_{\text{ext}},$$
$$D'_e = \varepsilon_1(i\omega\hat{n}\cdot A_1^{\text{ext}} - \phi'_1) - \varepsilon_2(i\omega\hat{n}\cdot A_2^{\text{ext}} - \phi'_2),$$
$$\vec{\alpha} = \vec{\alpha}' - \Sigma_1\Delta A_{\text{ext}} + i\omega\hat{n}L_1\Delta\phi_{\text{ext}},$$
$$\vec{\alpha}' = \Delta A'_{\text{ext}} + i\omega\hat{n}(\mu_1\varepsilon_1\phi_1 - \mu_2\varepsilon_2\phi_2),$$
$$\Sigma = \Sigma_1 L_1 - \Sigma_2 L_2 + \omega^2\hat{n}\cdot(L_1 - L_2)\Delta^{-1}\hat{n}(L_1 - L_2), \quad\text{and}\quad \Delta = \Sigma_1 - \Sigma_2.$$

---

[2]Note the different unit system in [5].

A.4   Spherical Harmonics

Many problems in physics lead to equations containing the Laplace operator $\triangle$, e.g. the wave or the Poisson equation. Depending on the symmetry of the considered problem one will try to solve the equations with a suitable set of variables. If spherical symmetry is present, one will choose the spherical coordinates (r,θ,φ). For example the Laplace equation

$$\triangle f(r, \theta, \varphi) = 0 \tag{A.12}$$

can be solved by the ansatz $f(r,\theta,\varphi) = R(r)S(\theta)T(\varphi)$ (see e.g. [6]). The Laplace operator in this coordinate system reads

$$\triangle \equiv \left[\frac{1}{r^2} \frac{\partial}{\partial r} \left(r^2 \frac{\partial}{\partial r} \right) + \frac{1}{r^2 \sin \theta} \frac{\partial}{\partial \theta} \left(\sin \theta \frac{\partial}{\partial \theta} \right) + \frac{1}{r^2 \sin^2 \theta} \frac{\partial^2}{\partial \varphi^2} \right], \tag{A.13}$$

and the above product ansatz allows us to separate Eq. (A.12) into three parts. The expression for φ yields a wave equation with the solutions $\{e^{im\varphi}, e^{-im\varphi}\}$, and the one for the variable θ has the form of an associated Legendre differential equation with the associated Legendre polynomials $P_l^m(\cos \theta)$ [7] as solution. Therefore the angular part of the complete solution is given by

$$T(\varphi)S(\theta) = e^{\pm im\varphi} P_l^m(\cos \theta). \tag{A.14}$$

This leads to the definition of an orthogonal system of functions, and together with a normalization factor[3] the spherical harmonics Y_{lm} [7] are derived[4]:

$$Y_{lm}(\theta, \varphi) = \sqrt{\frac{(2l+1)}{4\pi} \frac{(l-m)!}{(l+m)!}} P_l^m(\cos \theta) e^{im\varphi}. \tag{A.15}$$

A detailed explanation and rigorous definition of these functions can be found in the book of Varshalovich et al. [8], which we will follow from here on.

In Eq. (A.15) we see that a spherical harmonic $Y_{lm}(\theta,\varphi)$ is a single-valued, continuous, bounded complex function of two real arguments θ,φ with $0 \leq \theta \leq \pi$ and $0 \leq \varphi < 2\pi$. For a given l there exist $(2l + 1)$ functions corresponding to different m's.

---

[3]Unfortunately also different normalization factors and signs can be found in the literature. We will stick to the definition (A.15).

[4]Another less obvious but more elegant way is the definition of spherical harmonics as components of some irreducible tensor of rank l with the commutation relations $[L_\mu, Y_{lm}(\theta, \varphi)] = \sqrt{l(l+1)} C_{lm-\mu}^{lm+\mu} Y_{lm+\mu}(\theta, \varphi)$ where $L_\mu(\theta, \varphi)$ is a spherical component of the operator $\boldsymbol{L}$. See [8] for further details.

These functions are eigenfunctions of the orbital angular momentum operator in quantum mechanics which is defined as ($\hbar = 1$) [9]

$$L \equiv r \times p \equiv \frac{1}{i} (r \times \nabla).$$ (A.16)

The eigenvalue equations read

$$L^2 Y_{lm}(\theta, \varphi) = l(l+1) Y_{lm}(\theta, \varphi),$$ (A.17)

$$L_z Y_{lm}(\theta, \varphi) = m Y_{lm}(\theta, \varphi),$$ (A.18)

or in expanded form

$$\left[\frac{1}{\sin \theta} \frac{\partial}{\partial \theta} \left(\sin \theta \frac{\partial}{\partial \theta} \right) + \frac{1}{\sin^2 \theta} \frac{\partial^2}{\partial \varphi^2} + l(l+1) \right] Y_{lm}(\theta, \varphi) = 0,$$ (A.19)

$$\left[i \frac{\partial}{\partial \varphi} + m \right] Y_{lm}(\theta, \varphi) = 0.$$ (A.20)

In Eq. (A.17) one can see that l specifies the absolute value of orbital angular momentum $\left[$because $l(l+1)$ is the eigenvalue of $L^2\right]$ and m in Eq. (A.18) is the eigenvalue of L_z, which is the projection of the orbital angular momentum operator on the quantization axis.

Equation (A.19) has two linearly independent solutions for fixed l and m, but only one of them is regular $\left(\text{i.e. satisfies the condition } |Y_{lm}(\theta, \varphi)|^2 < \infty \right)$ while the other is singular at $\theta = 0$ and $\theta = \pi$. In quantum mechanics and electrodynamics the regular solution is of major interest. The homogeneous boundary conditions

$$Y_{lm}(\theta, \varphi \pm 2\pi n) = Y_{lm}(\theta, \varphi),$$ (A.21)

$$\frac{\partial}{\partial \varphi} Y_{lm}(\theta, \varphi)\big|_{\theta=0,\pi} = 0,$$ (A.22)

lead to integer values of l and m (with $|m| \leq l$). Since the differential equations (A.19) and (A.20) together with the boundary conditions determine the spherical harmonics only up to some arbitrary complex factor, a normalisation of the functions is required. With these tools at hand, we are able to expand an arbitrary function $f(\theta, \varphi)$ in series of spherical harmonics, provided that the function is defined in the interval $0 \leq \theta \leq \pi$, $0 \leq \varphi < 2\pi$ and satisfies the condition

$$\int_0^{2\pi} d\varphi \int_0^{\pi} d\theta \sin \theta \, |f(\theta, \varphi)|^2 < \infty.$$ (A.23)

The expansion then yields as

$$f(\theta, \varphi) = \sum_{l=0}^{\infty} \sum_{m=-l}^{l} a_{lm} Y_{lm}(\theta, \varphi), \tag{A.24}$$

with the expansion coefficients a_{lm} given by

$$a_{lm} = \int_{0}^{2\pi} d\varphi \int_{0}^{\pi} d\theta \, \sin \theta \, Y_{lm}^{*}(\theta, \varphi) f(\theta, \varphi). \tag{A.25}$$

This last relation may be considered as an integral transformation of $f(\theta, \varphi)$ from the continuous variables θ, φ to the discrete variables l, m. The transformation matrix in this case is given by $Y_{lm}(\theta, \varphi) \equiv \langle \theta, \varphi | lm \rangle$[5]:

$$\langle lm | f \rangle = \langle lm | \theta, \varphi \rangle \langle \theta, \varphi | f \rangle, \tag{A.26}$$

where

$$\langle lm | f \rangle \equiv a_{lm}, \qquad \langle lm | \theta, \varphi \rangle \equiv Y_{lm}^{*}(\theta, \varphi), \qquad \langle \theta, \varphi | f \rangle \equiv f(\theta, \varphi). \tag{A.27}$$

The expansion (A.24) is widely used in different branches of physics. It is called the *multipole expansion* and a_{lm} are the *multipole moments* (e.g. see [10]). The transformations (A.25) and (A.26), respectively, are unitary

$$\langle f | lm \rangle \langle lm | f \rangle = \langle f | \theta, \varphi \rangle \langle \theta, \varphi | f \rangle, \tag{A.28}$$

and the expansion coefficients a_{lm} satisfy the *Parseval condition* (see [6] or [7])

$$\sum_{l=0}^{\infty} \sum_{m=-l}^{+l} |a_{lm}|^2 = \int_{0}^{2\pi} d\varphi \int_{0}^{\pi} d\theta \, \sin \theta \, |f(\theta, \varphi)|^2. \tag{A.29}$$

A.4.1   Generating Functions for the Vector Harmonics

In Sect. 3.2 and Eq. (3.31) we see, that the time-harmonic electromagnetic field in a linear, isotropic, homogeneous medium fulfills a wave equation:

$$(\nabla^2 + k^2) \begin{Bmatrix} E(r, \omega) \\ B(r, \omega) \end{Bmatrix} = 0 \, .$$

---

[5]Summation or integration is assumed over all variables which are repeated twice.

By following [11] let us now construct a vector $\mathcal{M}$ with a given scalar function ψ and an arbitrary constant vector c:

$$\mathcal{M} = \nabla \times (c\psi). \tag{A.30}$$

Note that this vector is always perpendicular to c: $\mathcal{M} = -c \times \nabla\psi$. Since the divergence of the curl of any vector function vanishes, we immediately get

$$\nabla \cdot \mathcal{M} = 0. \tag{A.31}$$

If we insert $\mathcal{M}$ into the wave equation and use the following identities

$$\nabla \times (A \times B) = A(\nabla \cdot B) - B(\nabla \cdot A) + (B \cdot \nabla)A - (A \cdot \nabla)B,$$

$$\nabla(A \cdot B) = A \times (\nabla \times B) + B \times (\nabla \times A) + (B \cdot \nabla)A + (A \cdot \nabla)B,$$

we see that

$$\nabla^2\mathcal{M} + k^2\mathcal{M} = \nabla \times [c(\nabla^2\psi + k^2\psi)]. \tag{A.33}$$

If now ψ is a solution to the scalar wave equation

$$\nabla^2\psi + k^2\psi = 0, \tag{A.34}$$

$\mathcal{M}$ automatically fulfills the vector wave equation.

References

1. P.J. Mohr, B.N. Taylor, D.B. Newell, CODATA recommended values of the fundamental physical constants: 2010. Rev. Mod. Phys. **84**, 1527 (2012).
2. E.D. Palik (ed.) *Handbook of Optical Constants of Solids* (Academic Press, New York, 1985). ISBN 978-0125444200
3. T.J. Bruno, P.D.N. Svoronos, *CRC Handbook of Fundamental Spectroscopic Correlation Charts* CRC Press (Taylor & Francis Group, 2006). ISBN 978-0849332500
4. C. Sönnichsen, Plasmons in metal nanostructures. Ph.D. thesis, Fakultät für Physik der Ludwig-Maximilians-Universität München, 2001
5. F.J. García de Abajo, A. Howie, Retarded field calculation of electron energy loss in inhomogeneous dielectrics. Phys. Rev. B **65**, 115418 (2002).
6. C.B. Lang, N. Pucker, *Mathematische Methoden in der Physik* (Spektrum Akademischer Verlag, Heidelberg, Berlin, 2010). ISBN 978-3-8274-1558-5
7. M. Abramowitz, I.A. Stegun, *Handbook of Mathematical Functions* (Dover Publications, New York, 1973). ISBN 978-0486612720
8. D.A. Varshalovich, A.N. Moskalev, V.K. Khersonskii, *Quantum Theory of Angular Momentum* (World Scientific, River Edge, 1988). ISBN 978-9971509965

9. J.J. Sakurai, *Modern Quantum Mechanics* (Addison-Wesley Publishing, Reading, 1994). ISBN 978-0201539295

10. J.D. Jackson, *Classical Electrodynamics* (Wiley, New York, 1962). ISBN 978-0-471-30932-1

11. C.F. Bohren, D.R. Huffman, *Absorption and Scattering of Light by Small Particles* (Wiley-Interscience, New York, 1983). ISBN 978-0471293408

Appendix B
MATLAB Script for Mie Solution

With the following code example the scattering, absorption and extinction cross section for a spherical particle of arbitrary size can be calculated (the MATLAB® files are also contained in the MNPBEM toolbox [1–3]). The dielectric data for the sphere has to be provided in an ASCII table containing three columns of the form ω $\tilde{n}$ $\tilde{k}$, viz. [4, 5].

```
1  %   MATLAB code for Mie solution of Maxwell's equations.
2  %
3  %     written by U. Hohenester and A. Truegler
4  %     Karl-Franzens-University Graz, Austria
5  %     contact: ulrich.hohenester@uni-graz.at
6
7  %  diameter of sphere in nm
8  diameter = 50;
9  %  energy range in nm
10 enei = linspace( 300, 800, 201 );
11
12 %  background dielectric function and refractive index
13 nb = 1.34; epsb  = nb ^ 2;
14 %  dielectric function of particle (e.g. data from Johnson
15 %  & Christy)
16 epsin = epstab( enei, 'gold.dat' );
17 %  ratio of dielectric functions
18 epsz = epsin / epsb;
19
20 %  wavevector of light
21 k = 2 * pi ./ enei * nb;
22
23 %  preallocate memory for cross sections
24 scac = zeros( size( enei ) );   extc = zeros( size( enei ) );
25
26 %  table of spherical harmonic degrees, lmax = max. number of
27 %  degrees
28 ltab = []; lmax = 50;
```

© Springer International Publishing Switzerland 2016

A. Trügler, *Optical Properties of Metallic Nanoparticles*, Springer Series in Materials Science 232, DOI 10.1007/978-3-319-25074-8

```
29  for j = 1 : lmax
30    ltab = [ ltab; j * ones( 3, 1 ) ];
31  end
32  [ l, ind ] = unique( ltab );  %  find unique values
33
34  %  calculate scattering and extinction cross section
35  for i = 1 : length( enei )
36    %  Mie coefficients
37    [ a, b ] = miecoefficients( k( i ), diameter, ...
38        epsz( i ), 1, ltab );
39    %  scattering cross section
40    scac( i ) = 2 * pi / k( i ) ^ 2 *  ...
41        ( 2 * l' + 1 ) * ( abs( a( ind ) ) .^ 2 + ...
42        abs( b( ind ) ) .^ 2 );
43    %  extinction cross section
44    extc( i ) = 2 * pi / k( i ) ^ 2 *  ...
45        ( 2 * l' + 1 ) * real( a( ind ) + b( ind ) );
46  end
47
48  %  absorption cross section
49  absc = extc - scac;
50
51  %  plot results
52  plot( enei, extc, 'r.-', enei, absc, 'b.-', enei, scac,
53      'm.-')
54  %  annotate plot
55  title( 'Mie solution for spherical particle', ...
56      'FontSize', 18, 'FontWeight', 'b' );
57  xlabel( 'Photon wavelength (nm)', ...
58      'FontSize', 16, 'FontWeight', 'b' );
59  ylabel( 'Cross sections', ...
60      'FontSize', 16, 'FontWeight', 'b' );
61
62  set( gca, 'FontSize', 14 );
63  legend( 'extinction', 'absorption', 'scattering' );
```

The function epstab.m reads in the tabulated data from an ASCII file and interpolates for missing wavelengths.

```
1   function eps = epstab( w, fdata )
2   %  EPSTAB  interpolate dielectric function for
3   %    tabulated values.
4   %    Written by U. Hohenester and A. Truegler,
5   %    Karl-Franzens-University Graz, Austria
6   %
7   %  INPUT
8   %    w     ...  photon wavelength
9   %    fdata ...  file that stores dielectric function in
10  %               the form ene n k, with energy ENE in eV,
11  %               real and imaginary parts N and K of the
12  %               refractive index
13  %
```

```
14  %  OUTPUT
15  %    eps    ...  interpolated values of dielectric function
16
17  %  read in tabulated data
18  [ ene, n, k ] = textread( fdata, '%f %f %f', ...
19      'commentstyle', 'matlab' );
20
21  %  change energies from eV to nm
22  enei = (1 / 8.0655477e-4) ./ ene;
23
24  %  spline for interpolation
25  ni = spline( enei, n );
26  ki = spline( enei, k );
27
28  %  calculate dielectric function for wavelength w
29  eps = ( ppval( ni, w ) + 1i * ppval( ki, w ) ) .^ 2;
```

The function `miecoefficients.m` calculates the Mie coefficients of Eq. (4.10).

```
1   function [ a, b, c, d ] = miecoefficients( k, diameter, ...
2       epsr, mur, l )
3   %  MIECOEFFICIENTS - Mie coeff. according to Bohren and
4   %    Huffman (1983).
5   %
6   %  Input
7   %    k          : wavevector of light outside of sphere
8   %    diameter   : diameter of sphere
9   %    epsr       : dielectric constant of sphere
10  %    mur        : magnetic permeability of sphere
11  %    l          : angular momentum components
12  %  Output
13  %    Mie coefficients a, b, c, d
14
15  %  refractive index
16  nr = sqrt( epsr * mur );
17
18  %  compute Riccati-Bessel functions
19  [ j1, h1, zjp1, zhp1 ] = riccatibessel( nr * k * ...
20      diameter / 2, l );
21  [ j2, h2, zjp2, zhp2 ] = riccatibessel(     k * ...
22      diameter / 2, l );
23
24  %  Mie coefficients for outside field
25  a = ( nr ^ 2 * j1 .* zjp2 - mur * j2 .* zjp1 ) ./ ...
26      ( nr ^ 2 * j1 .* zhp2 - mur * h2 .* zjp1 );
27  b = ( mur * j1 .* zjp2 - j2 .* zjp1 ) ./ ...
28      ( mur * j1 .* zhp2 - h2 .* zjp1 );
29
30  %  Mie coefficients for inside field
31  c = ( mur * j2 .* zhp1 - mur * h2 .* zjp2 ) ./ ...
32      ( mur * j1 .* zhp2 -       h2 .* zjp1 );
33  d = ( mur * nr * j2 .* zhp2 - mur * nr * h2 .* zjp1 ) ./ ...
34      ( mur ^ 2  * j1 .* zhp1 - mur *      h2 .* zjp1 );
```

The function `riccatibessel.m` provides the spherical Bessel functions, see Eq. (4.11).

```matlab
1  function [ j, h, zjp, zhp ] = riccatibessel( z, ltab )
2  %  RICCATIBESSEL  Riccati-Bessel functions.
3  %     Abramowitz and Stegun, Handbook of Math. Functions,
4  %     Chap. 10.
5  %
6  %  INPUT
7  %    z      ... argument
8  %    ltab   ... angular momentum components
9  %
10 %  OUTPUT
11 %    j      ... spherical Bessel function of order 1
12 %    h      ... spherical Bessel function of order 2
13 %    zjp    ... [ z j( z ) ]'
14 %    zhp    ... [ z h( z ) ]'
15
16 %  unique angular component vector
17 l = 1 : max( ltab );
18
19 %  spherical Bessel functions of first and second kind,
20 %  see equations (10.1.1) , (10.1.11) , (10.1.12)
21 j0= sin( z )/z; j=sqrt( pi/( 2*z ) )*besselj( l(:)+0.5, z );
22 y0=-cos( z )/z; y=sqrt( pi/( 2*z ) )*bessely( l(:)+0.5, z );
23
24 %  spherical Bessel function of third kind (10.1.1)
25 h0 = j0 + 1i * y0;  h = j + 1i * y;
26
27 %  derivatives (10.1.23), (10.1.24)
28 zjp = z * [ j0; j( 1 : length( l ) - 1 ) ] - l( : ) .* j;
29 zhp = z * [ h0; h( 1 : length( l ) - 1 ) ] - l( : ) .* h;
30
31 %  table assignment
32 j = j( ltab );    zjp = zjp( ltab );
33 h = h( ltab );    zhp = zhp( ltab );
```

References

1. U. Hohenester, A. Trügler, MNPBEM – a Matlab toolbox for the simulation of plasmonic nanoparticles. Comput. Phys. Commun. **183**, 370 (2012).
2. U. Hohenester, Simulating electron energy loss spectroscopy with the MNPBEM toolbox. Comput. Phys. Commun. **185**(3), 1177–1187 (2014).
3. J. Waxenegger, A. Trügler, U. Hohenester, Plasmonics simulations with the MNPBEM toolbox: consideration of substrates and layer structures. Comput. Phys. Commun. **193**, 138–150 (2015).
4. P.B. Johnson, R.W. Christy, Optical constants of the noble metals. Phys. Rev. B **6**, 12 (1972).
5. C.F. Bohren, D.R. Huffman, *Absorption and Scattering of Light by Small Particles* (Wiley-Interscience, New York, 1983). ISBN 978-0471293408

Appendix C
List of Equations

© Springer International Publishing Switzerland 2016
A. Trügler, *Optical Properties of Metallic Nanoparticles*, Springer Series
in Materials Science 232, DOI 10.1007/978-3-319-25074-8

List of Figures

© Springer International Publishing Switzerland 2016   203
A. Trügler, *Optical Properties of Metallic Nanoparticles*, Springer Series
in Materials Science 232, DOI 10.1007/978-3-319-25074-8

List of Tables

© Springer International Publishing Switzerland 2016
A. Trügler, *Optical Properties of Metallic Nanoparticles*, Springer Series
in Materials Science 232, DOI 10.1007/978-3-319-25074-8

Index

© Springer International Publishing Switzerland 2016
A. Trügler, *Optical Properties of Metallic Nanoparticles*, Springer Series
in Materials Science 232, DOI 10.1007/978-3-319-25074-8